GARFIELD BENJAMIN

# MISTRUST ISSUES

How Technology Discourses Quantify,
Extract and Legitimize Inequalities

BRISTOL
UNIVERSITY
PRESS

First published in Great Britain in 2023 by

Bristol University Press
University of Bristol
1-9 Old Park Hill
Bristol
BS2 8BB
UK
t: +44 (0)117 374 6645
e: bup-info@bristol.ac.uk

Details of international sales and distribution partners are available at
bristoluniversitypress.co.uk

British Library Cataloguing in Publication Data
A catalogue record for this book is available from the British Library

ISBN 978-1-5292-3087-1 hardcover
ISBN 978-1-5292-3088-8 ePub
ISBN 978-1-5292-3089-5 ePdf

Cover design: Blu Inc
Front cover image: Stocksy/Giada Canu
Bristol University Press use environmentally responsible
print partners.
Printed and bound in Great Britain by CPI Group (UK)
Ltd, Croydon, CR0 4YY

FSC
www.fsc.org
MIX
Paper | Supporting
responsible forestry
FSC® C013604

For $M^1L_0$ and $R^eR_0$

# Contents

# About the Author

Dr Garfield Benjamin is Senior Lecturer in Sociology at Solent University, where they research and teach social issues in technology. They previously worked at the Birmingham Centre for Cyber Security and Privacy, after gaining a PhD from the University of Wolverhampton (published as *The Cyborg Subject* by Palgrave Macmillan in 2016). Garfield's work focuses on the social narratives performed around platforms, privacy, algorithms, data and AI. They seek to examine and challenge the power structures in which sociotechnical imaginaries and assemblages are embedded. Garfield's work has appeared in *ACM FAccT*, *Internet Policy Review*, and *Surveillance & Society*, among others.

# Acknowledgements

This book was sort of an accident. What started as an idea noodled for a conference became an academic article, then needed to be two articles exploring specific contexts, then spiralled into what became this short book. I am grateful to Bristol University Press for their Shorts series that enables works like this to be published in an appropriate format and timeframe. I would like to thank all the staff at Bristol University Press, particularly Paul Stevens – no one could ask for a more supportive editor – and Georgina Bolwell for keeping the production on track. I would also like to thank the reviewers for giving their valuable time to providing motivating and constructive comments on the proposal and the manuscript draft. Their input definitely made the book stronger.

Though we might spend hours alone writing, academic work is never really a solo effort. I would like to thank all those at Solent University who have offered support and encouragement not only for this project but also my wider research. Particularly, I would like to thank Brian McDonough, Philippa Velija, Francesca Pierporan, Donna Peberdy and Karen Burnell; and many other colleagues as well. Thanks to Sofia Carpio and Mayra Flores for organizing the 'Data (Dis)trust in the Global South' panel at 4S 2021, for including the initial ideas of this work in the programme and providing a space to share and learn. I would also like to acknowledge all those whose work has inspired me: things to read, concepts to explore, pointers of things to look at, conversations, questions, comments, and all the other forms of interaction that makes academia function and fun. There are too many people to list here, but the reference list and my social media follows are good places to start.

I would like to acknowledge our tiny humans, Mischa and Rémi, for making the days wildly more chaotic and fun amid the pressures of preparing a manuscript, whether that is yelling at crows or a wide-ranging flow of constant conversation on the bus. Finally, and most importantly, I would like to thank the other Dr Benjamin for being a constant inspiration, as an academic and a human being, and showing me what it is like to accidentally write a book!

# ONE

# Introduction: Trust Issues

Technology suffers serious trust issues. Between citizens and governments, users and platforms, governments and tech companies, the media and everyone, mistrust is rife. With good reason. And yet, we still act as if we trust technologies, the people who make them, and the people who use them on us. Every day we still use untrustworthy technologies designed by untrustworthy companies or have them used on us by untrustworthy organizations. We act as if we trust, even when we do not. We are expected to perform the conditions of trust even as trust crumbles with every new leak or revelation about the problems caused by unjust uses of technology. The problem has become not only a lack of trust, but that the prevailing discourses and power structures that surround technology have defanged the potential of mistrust to create change. How is it that the conditions of trust are created even in its absence? How is it that the legitimacy that comes from trust is extracted from people and populations?

The ways we talk about technology emphasize the need for trust. This might be to encourage the adoption of new technologies, to justify expanding the use of existing technologies, to support further research, or to back up (or counter) regulation. This book details how trust is quantified and extracted from populations. This process – which I call trustification – works in a similar way to how consent is extracted from individuals when they click 'accept' on a cookie popup notification. It is a sleight of hand that legitimizes not only specific uses of technology but the wider set of power relations in which they are developed and deployed, often against minoritized groups. The stakes of trustification are high, extracting legitimacy across local and global scales, and exacerbating existing inequalities in the process.

I do not use trustification in the financial–legal sense, but in relation to issues of quantification, datafication and the other processes of technologizing society. By operationalizing trust, converting it into a metric to be achieved, trustification extracts legitimacy from people to further the aims of using technology to manage society. But trustification is also about discourses, the stories we tell about technology and the way trust is used by and for power. These stories themselves legitimize the framing of trust as a quantifiable and solvable problem.

Bringing in political understandings of trust and mistrust, I describe how trustification works – in general, and in the specific contexts of government, corporations, research and the media – to legitimize exploitation and the increasing of inequalities. An example is the power games at work in the Facebook Zero scheme, a deal with mobile data providers in over 50 countries that offers free access to Facebook (alongside specifically selected other websites) in exchange for embedding corporate infrastructure and data practices: trustification, control and market dominance couched in the language of humanitarianism and beneficence. The stories technology discourses tell us often use 'for good' rhetoric while eliding the question of what good or whose good, with trust standing in for accountability. But far from being a problem of misplaced trust, of blame placed on users and the used, I argue that forces of power actively extract and operationalize trust.

The central argument of this book is that technology discourses performatively manufacture trust and leverage that trust to fulfil these lopsided aims, while reducing the critical power of mistrust to counter inequalities. It is the act of measuring trust that creates it and the conditions of power and legitimacy they provide. Trust in these contexts tends to be used in a vague sense, to justify a range of priorities and to cover over motivations that often avoid explicit definition or shift over time. Trust is not built but extracted from populations for the purpose of justifying the legitimacy and continued dominance of certain ideas: perceptions of objectivity and faith in technology (Beer, 2018: 24–5); the seductive pull of quantifying people and complex social issues in response to uncertainty (Milan, 2020: 1); permissionless innovation and decontextualized solutionism (Morozov, 2014). In short, technological power 'trustifies' people to support the continued dominance of technology industries and technological thinking in society while avoiding proper scrutiny.

Trustification makes it even more difficult to address the underlying power structures and sociotechnical asymmetries that perpetuate and exacerbate inequalities and injustices. Through this book I argue that trustification has permeated the narratives of governments, corporations, research and the media. I offer a framing of trustification as a productive lens through which to examine the interwoven threads of power and discourse that enable the continued inequitable design, development and deployment of technologies.

## Technology and trust

In his 1984 Turing Award acceptance, titled 'Reflections on trusting trust', Ken Thompson claimed that 'you can't trust code that you did not totally create yourself' (1984: 3). While this may be true, and was a sustainable position among programmers in 1984, today it is simply untenable. Every day, every moment, we trust that code (and other technologies) will function as expected. Notably, however, Thompson also suggested 'perhaps it is more important to trust the people who wrote the software' (1984: 1). The shifting emphasis between trusting technologies versus trusting people who make, sell or use those technologies is an important one. And it bridges different loci of trust: trust in function and trust in motives. Maintaining these distinctions has become increasingly important with the way technologies are embedded throughout our lives and society.

How do we identify and maintain separation between the locations of trust when they are often spread across complex assemblages of technical objects, designers, users, subjects, regulators and narratives? What is a 'technology' in relation to trust? We often talk about technologies in metaphorical terms, which can be a useful aid to understanding and confronting complex issues. Placing a target on 'AI', for example, allows us to direct critique at a whole host of different tools, techniques, framings and practices. But this same use of language can conflate issues and occlude matters of power and inequality. If we are asked to trust a technology as an object, it can become difficult to understand or find who or what we are trusting. This is not a matter of philosophical musings about whether a technical object is an entity it is possible to trust. It is about situating that object in society, in a set of relations between different people.

Any given technology is a hybrid of constituent technologies. Just think of the specific hardware, operating system, libraries, data and other components that support an AI algorithm. A technology is also the product of human labour and decision making in designing and making all those constituent parts. A technology is also the product of a lengthy process of research and development, like the fields of AI or data science. A technology is also a politics, linking a specific purpose to a wider agenda. Think of the push for surveillance or the prioritization of business models in the data economy. A technology is also the story we tell about all these aspects, the discourses that surround specific technologies and the narrative of technology in general. Incorporating all these different things a technology can be, I will therefore be using a broad view of 'technology' as a relational sociotechnical assemblage. This includes technical objects but also considers the people, power structures and discourses that create, use and legitimize them.

Thinking about technology in this way, as a set of social relations, also helps emphasize the contexts within which it operates, the power relations and narratives it emerges from, entrenches and creates. It helps reinsert the different and unequal groups of people behind technologies: the people data is about, the people labouring to make the systems, the people designing them, the people using them, and the people making the decisions. It is important to consider how the discourses that surround assemblages of technology are incorporated into practices. Discourses shape and limit what questions can be asked, in ways that invite specific answers that reflect specific priorities. Technology discourses function by sliding trust from one type of entity (object, concept, infrastructure, individual, organization) onto another. They cover over the points where quantification would otherwise clearly fail – for example spreading the narrative that technological solutions always 'work' and therefore deserve our trust.

The idea that technologies like AI need and deserve trust heavily permeates regulatory and media discourses. For example, the EU High Level Expert Group on AI has repeatedly resolved different technical and social issues down to the need for trust. Having these assumptions underpinning critical work on future regulation furthers the sliding of motivations into technical functions. It contributes to the blurring of trust between entities in highly asymmetric power structures that privilege dominant economic and political narratives.

## Trust and power

What do I mean when I say 'trust'? In this book I rest my argument on a political understanding of trust (see, for example, Hooghe et al, 2017) to acknowledge the elusive nature of the concept and the way its use is always embedded in broader political, social and economic aims and agendas. This form of trust is multidimensional (Norris, 1999) and requires consideration of the loci of trust as much as its different types. Christina Boswell (2018) builds on standard definitions about expectations, intentions and actions by adding not only a condition of uncertainty but also that such expectation 'provides the basis for action' (2018: 34) and thereby places both responsibility and authority on the trustee.

Two observations by Russell Hardin (2006) further contribute to relevant issues of trust for my discussion. Firstly, the root of Hardin's account of trust is a push to think about trust as a derivative of trustworthiness. Taking such an approach shifts the emphasis from the one holding the trust onto the one in whom that trust is held, instantly drawing into question issues of extraction, of 'getting' trust. Secondly, the origins of the word trust referred to a specific role and being trustworthy in fulfilling that role as expected (Hardin, 2006: 3). This emphasizes context – being trusted in a particular role but not necessarily in others – but it also highlights the performative nature of trust. Trust is active in both directions. It is not only something done by the truster. It is also performed by the trustee, combining roles of trust and trustworthiness. This quickly becomes a complex and messy situation, and 'the formation of trust attitudes is therefore relational, multi-faceted, and contextually embedded' (Eder et al, 2014: 4). Trust relations are power relations.

Hardin (2006: 17–18) describes how three main conceptions of trust – encapsulated interest, moral commitments and character disposition – combine knowledge and belief, an action that constitutes the relationship of trust. Hardin focuses on encapsulated interests – the trustee taking on the truster's interests as separate from their own, rather than interests simply aligning between truster and trustee. For example, an elected representative might encapsulate the interests of those they represent, regardless of whether their own interests agree. This form of trust includes reputation as a replacement for direct action and direct relationships.

I argue that technology discourses perform moral commitments and character disposition to sustain this speculative reputation.

Technology companies often ask for trust (and funding) on faith as they develop new systems with unknown, unintended or unwanted consequences. The 'for good' narrative re-emerges here. Rather than encapsulating people's specific interests, trust is formed in relation to vague notions of humanity, progress and convenience without problematizing whose good and whose interests are being encapsulated. This is further complicated by the displacing of trust onto technical objects, replacing even reputation with technical jargon and manipulative marketing.

Trust both exercises and creates power. This is necessarily circular. Part of the point is that trust is an elusive concept, a sliding signifier for the legitimacy of power, and never begins from a neutral set of social and power relations. The idea that trust starts from a neutral, informed decision echoes narratives that claim the neutrality of technology. This has a depoliticizing effect, shutting down criticisms that try to situate and contextualize technologies in broader sociotechnical assemblages. The idea that either technology or trust are value-free instils a focus on numerical values, which is itself not a neutral proposition in the normative role of quantification in framing social issues.

An understanding of trust must therefore always be counterbalanced by a productive mistrust – which is key to democratic accountability – and my argument builds on Zuckerman's (2021) emphasis on the transformational role of mistrust. The issue is not really a lack of trust but the operationalization of trust that removes the power of mistrust. Mistrust engages the affective, emotional, personal and interpersonal dimensions of trust as well as the systemic inequalities and injustices that underpin and are sustained by trust in power and technology. Trustification can be seen as an attempt to tip the scales, to abandon the reciprocal relation between trust and mistrust by metricizing trust as an achievable goal rather than as an ongoing set of complex social relations.

A dominant approach to the problem of trust has been to logic it away, and this underpins another key concern of this book. For example, the Trusted Platform Module acts as a separate processor with a limited (and therefore hypothetically secure) set of operations, an additional piece of hardware supposed to add trust to an otherwise untrustworthy device. Another example of computing logics applied to trust is the pursuit of 'trustless' systems, meaning systems that 'do not require trust'. But in wider usage, thinking about the relationships between people through and around technology as being trustless also carries implications of being 'untrustworthy'. It also risks turning the issue of trust round as a responsibility of the truster rather than something to be earned or built in a relational way.

The flipping of responsibility is a common trend in technology discourse and another key aspect of trust to approach critically: the impression that technology could solve all humanity's problems if only people trusted it. For example, there is often talk of 'trust deficits' in the use of technology, as if this is a problem with people rather than a problem with technologies and institutions. This has been particularly visible in 'solutions', policies and information campaigns during the COVID-19 pandemic.

Christina Boswell demonstrates the historical emergence of 'technocratic public management tools in the form of targets and performance measurements' (2018: vii) as an attempt to manufacture trust in political institutions in order to 'reground political trust' (2018: 178). These measurements create abstract images of social contexts in ways that support specific policy framings, leading to specific policy responses. Boswell demonstrates the failures of measuring trust in government, leading to unintended, and counterproductive, consequences of cynicism and mistrust. Her emphasis on narratives and the social construction of these policy framings around particular interests aligns well with the argument here: that technology discourses manipulate metrics to manufacture an abstracted concept of trust.

Another concept that is particularly relevant in this respect is the idea of proxy variables – quantifiable stand-ins for complex, inaccessible or unmeasurable issues – linked to questions of who or what is the object of trust (Devine et al, 2020). The issue of who needs trust, and from whom, is tied to issues of power. Confronting the power behind and from trust requires us to consider state, corporate, epistemic, colonial and mediated forms of power, and how they relate to one another in shaping the role of technology in controlling society.

Therefore, while different understandings or definitions of trust help inform and frame the argument developed in this book, I am not so much concerned with what trust *is*, and more with what it *stands in for* and what stands in for it. I am concerned with how trust is *used* to support various agendas and systems of power. I have already stressed the importance of the stories we tell around technologies, the narratives and discourses that constitute and legitimize our design, use and understanding of specific technical objects embedded in sprawling sociotechnical systems.

## Performative trust in technology

To analyse these issues of social narratives, I use a performative theory of technology. Initially outlined in relation to privacy (Benjamin, 2020) but applicable more widely, this theoretical position applies queer theorist Judith Butler's conception of performativity (1990; 2018) to confront the ways that individual and collective actions create and are created from wider narratives and norms of technology. When a role or concept is performed, it expresses existing societal norms and pressures, while these social expectations are also created through the collection of many actions performed by many people. This approach is useful to assess the roles we are expected to perform in relation to trust, and the influence of discourse in creating conditions for material practices. These roles and acts are repeated over time as embodiments of dominant narratives (such as states reflecting corporate interests and entrenching them in legislation, or media normalizing underlying assumptions even when they critique specific symptoms). The root of trust lies not in a logical decision starting from a detached neutral position, but in a whole host of social expectations and wider power structures.

The performative way that trust is constituted in technology discourses returns to the point with which I started this book. We often do not trust technology (as an object, social system or discourse). Against the hype is a 'growing resistance towards harmful algorithms and irresponsible data practices […] Public trust in AI is on rocky ground' (Peppin, 2021: 195). I argue that a lack of trust is not negative. Mistrust can and should exist alongside trust as an 'appropriate' response when, for example, 'data practices are not deemed trustworthy, as in the case of scandals about data breaches' (Kennedy et al, 2022, 395). We often cannot even tell who, what or where we are expected to place our trust in the power structures and sociotechnical assemblages that surround technology. This raises the need to focus on situated power rather than ethical distinctions: we cannot trust a complex sociotechnical system; the sliding of trust is counter to demonstrating trustworthiness; and trust is impossible or unhelpful to measure. Despite this, or even because it is disappearing as an appropriate value to rely on for legitimacy, trust is increasingly measured.

The quantification of trust in its very absence is where performative norms come in. Despite the lack or impossibility of trustworthiness, we are still expected to trust technology, we still act as if we do trust it. With each act of compliance, and each utterance of the discourse that expects and normalizes it, the conditions of trust are met anyway. The specific object or system becomes a proxy for a precise measurement of trust in a vague narrative. The embedding of technology into every part of our individual and social lives normalizes these expectations. We are expected to use technology regardless of whether we trust it, and with every act that follows this expectation, we contribute to performing the norm. It is a circular justification in which quantified values of trust become an empty signifier for mistrust. In the process, mistrust loses its critical and democratic potential, and accountability becomes meaningless.

Measuring trust therefore extracts legitimacy only through the act of measurement, supported by the normalized and enforced acts that perform the conditions of trust. These conditions provide technology discourses with the power to further the narrative of technology at all costs. The discussion in this book is therefore less concerned with defining trust in technology discourses, and more focused on picking apart the conditions of trust, forcing our way through the smoke and mirrors of trustification, and examining how discourses extract and operationalize trust as a proxy for legitimizing power and sustaining inequalities.

## Outline of the book

From within this conceptual context, Chapter Two outlines a new position of trustification. It does this by situating it in relation to several existing concepts and theories around technology and society: quantification or datafication of people and issues as a political process; gamification of activities into scores for motivation, action and control; and consentification as the manipulation of individualized agency to legitimize these practices. Mapping these processes in relation to one another leads towards trustification as the metricized extraction of legitimacy from populations, building on established critical perspectives and concepts to develop a new framework focused on how trust is operationalized according to persistent logics of computational power.

The following chapters each focus on a specific realm or logic of power, tracing the historical precursors of trustification into its performance through current technology discourses. In Chapter Three I examine state power through the measurement of authority and the ways that technologies of counting are used to surveille and control populations. Government and regulatory narratives are discussed in the extraction of trust in the state as a trade-off against other actors, and the spread of colonial logics is traced through the setting of legislation and standards. In Chapter Four I discuss corporate power and the historical development of capitalism and colonialism as twin logics of risk that measure trust in corporate structures and the technologies they produce. The spread of corporate cultures of quantification and trust lead into a discussion of innovation as a powerful narrative for justifying technological solutionism. In Chapter Five I shift the focus to research discourses, again examining the colonial roots of knowledge production and the imposition of measurement and quantification as a dominant global framework. I bring these epistemological histories forward into the contemporary university through metrics of academic success, as well as how the presentation of credibility trustifies populations and the ways that technology discourses shape national and global research funding agendas. In Chapter Six, the last of these chapters, I address media narratives and highlight the broader stories we tell around technology. This includes how technologies are represented and misrepresented to serve the aims of tech company executives, investors, politicians or other influential groups, as well as how platforms shape the spread of information in ways that quantify trust in information, technology, and the other forms of power that use them. Across these chapters, examples are drawn from uses of data, AI, platforms, and other forms of technology that permeate society in ways that extract and quantify trust and, in doing so, legitimize the perpetuation of inequalities and injustice.

I then provide three case studies that demonstrate the processes of trustification in the discourses surrounding specific technologies. These cut across state, corporate, research and media arenas, and examine the overlap and interaction between the different narratives and logics of power. I discuss COVID-19 tracing apps in the context of the UK and the

Global South. I interrogate the colonial agendas underlying 'tech for good' initiatives. I confront facial recognition and emotion recognition as embedding harmful narratives connecting different dimensions of marginalization. The case studies draw on examples referred to throughout the book but provide a more focused outline of how trustification is mobilized by different actors to extract legitimacy and perpetuate inequalities.

I conclude the book by demonstrating how trustification elides scrutiny by offering false trade-offs that shift discourse, visibility, attention and effort away from underlying issues of inequality and injustice. My argument returns to the issue of proxy variables to ask what is standing in for trust and what trust stands in for. I highlight the asymmetries between those giving trust and those who want it, stemming from problematic conceptions of trust as a measurable value. As a counter to these narratives, I end the book by restating an understanding of trust not as something we have but as something we do, and the role of narratives in shaping how decisions are made about technology. This opens a space for exposing the power structures behind trust and paths towards more reciprocal relationships building more equitable balances of trust and mistrust.

# TWO

# Trustification:
# Extracting Legitimacy

In this chapter, I outline the conceptual framework of trustification, a way of describing how legitimacy is extracted from populations through processes of quantification, creating the conditions of trust without the underlying social relations of trustworthiness. This is set out from the relation between technology and trust, through the processes at work, to the role of proxy variables that stand in for trust in legitimizing technology discourses. The socially constructed drive to technology, particularly the drive to AI or the drive to data, does not accept the interplay of trust and mistrust. It does not accept any challenge to objectivity narratives, any hint that problems are not 'solvable' (in principle or through technology). Such ambiguity does not fit within the discourses that surround the power and legitimacy of technology and those who design, build, sell or use it. Such a challenge to the absolute faith in technology as a force for good for individuals and society is not permitted by those discourses.

A nuanced, political, socially engaged, fluid approach to trust and mistrust is simply outside mainstream technology discourses because it challenges the power structures and inequalities that such narratives sustain. These narratives no longer seek to build trust, but to create the impression of it, to perform the conditions of it. Or, more bluntly, to extract it. This process of performative extraction between quantification and discursive power I call trustification.

## Trusting technology

People are asked every day to trust technology. This is in part trusting it to function as intended, which already involves not only expectations of operational accuracy (trust in the actions of the technology) but also expectations of intent (trust in both the intentions and actions of the designers or users of the technology). When we situate this operational form of trust across the sociotechnical assemblage that constitutes a given technology, we acknowledge other forms of trust as well. We are not only being asked to trust specific instances of specific technologies. We are not only being asked to trust the sprawling network of constituent technologies, materials and processes that make a specific technology. We are also being asked to trust in 'technology' as a broader concept and discourse, an organizing principle of society. This can be seen not only in trusting that our toaster will not catch fire, or our messaging app will not leak our private encrypted messages, or our traffic management systems will not malfunction to cause a crash. It also goes beyond trusting that the designers and operators have acted in ways that should prevent these failures from occurring. The sociotechnical assemblage includes social discourses (political, economic, cultural) that also expect to be trusted, and therefore gain legitimacy and power.

Influential organizations that set technical and legal standards taken up worldwide emphasize the need to trust technology: the WWW Consortium's mission for a 'web of trust' (W3C, 2012); the US standards agency NIST's 'trust and artificial intelligence' report (Stanton and Jensen, 2021); or the EU's High Level Expert Group on AI 'ethics guidelines for trustworthy artificial intelligence' (AI HLEG, 2019). The latter example appears to make progress towards emphasizing trustworthiness and thereby shifting the emphasis on demonstrating a reason to trust rather than placing the burden on the truster. However, it remains within the discourse of placing that trust in the technology itself, and thereby limits the scope of trust to functionality rather than embracing the broader political conceptions and implications of trust that should accompany any AI system that is automating decisions about people and society.

Mark Ryan (2020) focuses on the EU AI HLEG when describing trust in AI as a misplaced act of anthropomorphization, treating computers and other technical objects as people. Ryan highlights the normative and affective dimensions of trust. He places the lack of affective response as a reason AI cannot be treated as an entity that can be trusted – your WiFi-connected pacemaker does not care whether you trust the security of its communication protocols, it simply follows instructions that conform (regardless of their intention) – while also emphasizing the need to shift responsibility onto those designing systems. He identifies within the EU AI HLEG's description an acknowledgement that trust needs to continue throughout the entire sociotechnical context but responds critically to their decision to place the focus of trust on the AI itself anyway.

The NIST report also emphasizes trust as a human trait, but unlike Ryan it misses the point by continuing to direct that trust towards technologies without attending to the asymmetrical power dynamics and inequalities embedded within those technologies by humans. The importance of affect is another source of asymmetry. Someone trusting someone/-thing else is placing themselves in a position of vulnerability, and this carries not only material but also affective risk. Those designing or using the technologies on other people require that trust, but their stake is more related to power than affect. The impact on people's lives is subsumed under targets and the quest for power.

In this context, the framing of trust onto technical objects suggests a widespread (if unspoken) acknowledgement that those entities requiring trust (often governments or corporations) are untrustworthy. Displacing trust onto machines performs an imaginary trustworthiness based on function rather than vulnerability and politics, hiding untrustworthiness under the guise of technical competence. Shifting the locus of trust shifts the type of trust that is expected, obscuring power relations.

Assessing technologies as sprawling sociotechnical assemblages helps us to make those relations visible and thereby confront them and the power structures that maintain them. The issue at stake is not about a lack of trust or misplaced trust, it is about creating space for mistrust to emerge as a political and discursive tool for change.

## Technologizing trust

If trust is increasingly placed on technologies, it increasingly functions as a technology itself. According to the NIST report on trust and AI, 'trust serves as a mechanism for reducing complexity' (Stanton and Jensen, 2021). This raises the question of whether such trust is desirable. Reducing complexity always entails a loss of detail and accuracy. This reveals a tension within technology trust narratives, and within AI specifically, with increasingly complex systems that also seek to serve as a mechanism for reducing complexity. Implicit within the discourse of reports like NIST's (and the EU AI HLEG's) is a push to define trust as functionally analogous to AI. For, if AI serves the same function as trust, why would we not trust it? And the emphasis on trustworthiness being assigned to technologies, while loci of trust remain elusive within broader sociotechnical assemblages, furthers this displacement of trust.

There has been a longstanding narrative of trust as something that is measurable, quantifiable. Aside from the reduction of complexity and loss of context or nuance (trustworthiness and the value of mistrust), attempts to quantify trust are woven into a narrative whereby trust is a resource extracted from people rather than a set of relationships built and maintained. In discussing trust in numbers, Theodor Porter (1995) describes quantification as 'a technology of distance' and a 'strategy of communication' through which 'objectivity names a set of strategies for dealing with distance and distrust' (1995: ix). This supplanting of trust and mistrust with quantification is essential for placing trust in objects.

The relationship between a human and an object is, like trust, potentially highly affective and real to the human. But it is one-sided, reflecting trust in systems and power rather than interpersonal trust. The process of placing trust in technologies is therefore doubly a process of 'objectifying' trust: placing trust in an object; and converting that trust into objective measures. This trust is maintained only in so far as the technology continues to function as intended. The framings and narratives of measurability and objectivity therefore convert trust into a function of power. Echoing the ways social categories like race are constructed as technologies of power (Benjamin, 2019; Mbembe, 2019), this operationalization turns trust into a technology for the extraction of legitimacy and the exertion of power.

This gets to the heart of my argument in this book: trustification bypasses relational forms of trust, instead manufacturing a performance of trust through metricizing and operationalizing it. These logics of quantification and computation are entwined with discourses that support them and push solutionist narratives to sustain power and inequality. An individual, group or population might not trust you, but trustification steps in to mediate the legitimacy of power anyway.

Christina Boswell describes the problems of opacity, complexity and increasing specialization in contexts of political trust. The challenge then becomes not to directly build trust through relationships with individuals or publics, but instead to 'create conditions that enable people to bestow political authority under conditions of uncertainty', even in a state of disillusionment and the absence of trust (Boswell, 2018: 37). The challenge occurs even more so in the sprawling assemblages that surround any given technology or intervention, where overt political trust may be only one factor alongside technical, economic, reputational, social, cultural and narrative forms of trust. Technologized trust discourses attempt to collapse these complex sociotechnical assemblages onto specific technical objects or measures. We see here again the importance of discourse, of performing trustworthiness to manufacture not trust but the conditions and functions of trust.

Embedded within this performative discourse of trust is the assumption that trust can be measured. This assumption also becomes the goal in a performative self-fulfilling loop. Measuring trust has been described as a 'great lacuna' of social and political research (Glaeser et al, 2000: 811), something that, 'with some notable exceptions [...] does not have a long tradition' (OECD, 2017: 3). Trust has been increasingly measured as it has declined, and as technological tools have become complex enough to sustain the illusion.

Yet measurement has become the basis for asserting the legitimacy of political and technological trust in society, and the basis upon which trust provides legitimacy for political and technological action. Measurement in general has a strongly normative and performative element, where 'indicators of quality are taken as definitions of quality' (Biesta, 2017: 3). Definitions of trust are often absent from its measurement, particularly in population survey questions that reduce complex issues into quick numbers. In other words, in the manufacture of trust through the process of measurement and its outcome, creating the conditions for legitimacy creates that legitimacy.

This enacts a turning outwards of internal corporate/ technology discourses of performance measurement (and performing measurement). The context shift applies the same logics of research, product development and market values to publics from whom trust in complex sociotechnical assemblages is required. The understanding of trustification developed here in relation to technology discourses is rooted in computational logics that seek to convert social relations into computational models, which are perpetuated in computational narratives of trust as something that can be measured, solved or abstract away.

Trustification proxies for various other terms and processes of quantification. Categorization, ranking, targets, are all part of the way trustification operationalizes trust. All may stand in as proxies for building or demonstrating trust.

All are wrapped in the same discourses that legitimize the expansion of technological narratives, logics and interests that perpetuate affective and social harms of technologies. Trustification surrounds concepts like ranking, legitimizing them through the specific extraction of measurements that stand in for trust and the discourses that support this. I will outline how trustification emerges as a process in relation to, and as the convergence and expansion of, several existing '-ification' processes: datafication (the quantification of individuals); gamification (the quantification of populations); consentification (the extraction of legitimacy from individuals); and, finally, trustification (the extraction of legitimacy from populations). Throughout this tracing, I will use two examples to demonstrate the way these processes work in relation to trust: menstrual cycle tracking apps; and Amazon's Alexa voice assistant proposing to use voices based on the vocal patterns of customers' deceased relatives.

## Datafication

Datafication, which emerged as a term following the rise in discourses surrounding 'Big Data' (Mayer-Schönberger and Cukier, 2013: 73f), is rooted in the desire to quantify the world in order to understand it and influence it, from the natural sciences to the social sciences and beyond into the ever-expanding realm of computational sciences. These aims of measuring, replicating and predicting the world expanded, digitized and became one of the dominant logics of organizing society. Technology discourses cemented datafication as the process through which 'large domains of human life became susceptible to being processed via forms of analysis that could be automated on a large-scale' (Mejias and Couldry, 2019). This history of quantification and datafication has strong links to colonialism and commodification (Thatcher et al, 2016) and, returning to our focus on the construction of narratives, it emerges as a fiction in the creation of data (Dourish and Gómez

Cruz, 2018). José Van Dijk (2014) famously problematized the ideological component of datafication, and Salome Viljoen (2021) has added that it is 'a social process, not a personal one'. While it is the conversion of individual people (or parts of them) into representations in data, what we can describe as the *objectification of individuals*, datafication is a social, political and discursive choice made to convert masses of individuals into objects to enable *the manipulation of individuals as data*.

Ian Hacking (1990) described quantification as a style of reasoning that underpins contemporary thought, and this statistical and predictive view of the world and society embedded a drive to economic thinking. This is explored further by Jathan Sadowski under the name of data capitalism (2019), comparing the extraction and accumulation aspects of datafication to previous processes of financialization. They undertake similar approaches to using power and influence, and a similar imperative towards more capital, more surplus value, more data.

David Beer (2018: 24–5) describes how trust in numbers (see also Porter, 1995) and the derived trust in insights are linked to narratives built on a perception of and faith in objectivity. Importantly, from the process of building trust we see both 'how authority is given to data' and 'how expertise is demarcated' (Beer, 2018: 134), the creation of legitimacy for the data itself and for those using it. This echoes one of the central principles of this book: that technology discourses shift trust through sociotechnical assemblages, separating those 'in the know' from those assumed to trust them as part of legitimizing data.

Van Dijk (2014) similarly highlighted the links between datafication and trust. Social media relies on assumed trust in corporate platforms. This is carried through in practice (via infrastructures and metadata) and in method (via initiatives and policies) across further private and public organizations (from government departments to law enforcement to academia). Throughout the rest of this book, I examine the way trustification operates across these different contexts of an often tightly knit and self-replicating ideology of datafication and the data ecosystem.

Looking at political trust more widely, Boswell showed how data-driven targets worked as 'both signalling and disciplining devices' that 'can be analysed in terms of their capacity to produce trust, whether through imposing rules backed by sanctions (disciplining), or through establishing reliable truth claims (signalling)' (2018: 74). And yet these targets carry risk, particularly in public and political contexts where they often fail, leading in some cases to the need to further manipulate – twist, contort, spin – data to create the semblance of success upon which trust, and legitimacy, must rest. The outcome of this process, including the harms of allocation and erasure it inflicts, reflects data violence (Hoffmann, 2018), a form of administrative and discursive violence.

According to Dean Spade (2015), administrative violence is enacted through metricization and centralized goals. It is the administration of power and the imposition of a demand by power from those subjected to it. These systems of meaning and control mediate social relations and individuals' ability to know themselves through these discourses. It is the establishing of certain values appearing as neutral that enact unjust outcomes on individuals' lives (according to socially constructed categories such as race, gender, sexuality, nationality, disability, class and others).

In response, Spade outlines a critical trans politics built on Black feminist refusal, that seeks to shift the focus from individuals to systemic issues. Categorization is a key part of the manipulation of individuals as data, an allocative harm that can be seen in four processes of administrative violence: 'imposing an identity rejected by the concerned persons, the denial of socio-economic rights, a symbolic process of stigmatisation, a complete lack of transparency' (Beaugrand, 2011: 236), all justified through delegitimizing certain groups and legitimizing the power of those currently wielding it.

Power in technology discourses is performed by perpetuating norms of datafication and expectations that datafication can and should happen. These norms assign roles of data subjects who can and should be datafied, and data owners, processors and controllers who can and should define these processes. Trust specifically is measured, datafied, in ways that assign further categories of relative compliance, of satisfaction, of privilege.

To give examples, datafication in menstrual cycle tracking apps converts individuals' bodily functions into data to be monitored. This is based on assumptions of trust in that process of conversion, the translation of lived bodily experiences into numbers that can be tracked over time and averaged to offer predictions. Meanwhile, beyond the wider issues of rampant data collection with Amazon Alexa, the initiative to inspire trust by using the voice imprint collected from a deceased relative shows how the datafication of one person can be used to manipulate another through operationalizing the affective social relations between people to proxy an emotional connection to a network of technical systems.

## Gamification

Many of the dominant conceptions of trust within technology discourses, as in policy, media and research more generally, come from a combination of international relations and/or economics, particularly the legacies of the likes of Nash and Schumpeter. Fundamentally, these are built on conceptions of trust that are defined by contexts of negotiation, usually between countries or companies. Even within technology-focused issues such as the use of drones, these are often seen as issues of trust between governments rather than a relation between power and its victims. These contexts are often essentially competitive, and the theories behind them tend to operationalize trust, reverting to conflict and zero-sum thinking. But public trust is not zero sum, and it should not be competitive. It is certainly not a negotiation. And we must keep at the forefront of our thinking the massive power asymmetries that shape these contexts. In practice, matters of trust never start from a neutral power relation. They are always embedded in sociotechnical assemblages defined by power and discourse. The conversion of society into data converts societal problems into winnable games. Trust is gamified in these settings, converted into a single measure of risk as a proxy for complex social relations.

Gamification echoes the economization of discourse around public trust, and around technology discourses more widely in which efficiency narratives dominate social issues. It appears in the automation or offloading of politics by this imposition of economic approaches (Andrejevic, 2021). It is not so much the experiences of individuals that are gamified, although the design practices like ever-shifting scores, targets and achievements certainly contribute towards manipulating individuals. The discursive power of gamification is the imposition of a computational logic of governance. This operates through comparison. Gamification, whether of language learning, exercise or job performance, works by comparing individuals to one another. Leaderboards and the mediated sharing of scores again insert competition, reducing individuals to a measurable point on a scale against others. This divides the collective potential for mistrust of the gamified system by obscuring and distracting from the top-down control that sees individuals as datapoints to be manipulated.

Critical perspectives on gamification are therefore useful to inform our understanding of the gaming of trust in particular. Gamification has been described as a 'technology of government' (Whitson, 2015: 341, 339), a 'tool for Foucauldian biopolitics' (Schrape, 2014: 37) in which 'players become pawns in larger games of states and terrorists' (Whitson and Simon, 2014: 312). While gamification has been used to motivate individuals to do any number of things (higher performance in a workplace, learning a new language, increasing exercise), it is ambivalent about those individuals as individuals. Individual people are not the players in gamification.

At best, people are treated as NPCs (non–player characters) for wider games of promoting gamification as an end in itself. At worst, they are treated as mere resources for the wider games between organizations. As O'Donnell points out, 'these new "gamed" experiences enter into new dialogs with existing forms of power and structure' (2014: 357). Gamification is datafication as the governance of populations. The game that is being played is between rival companies for increased market share, or between organizations and some abstract ideal or target, like governments and policy goals towards public health, wellbeing, or environmental targets.

The ambivalence of gamification towards individuals is seen in the particular way it converts complex social issues into data. It is no longer even attempting to translate each individual into data. Gamification logics only require parts of individuals to be converted into data, for the value comes from these parts brought together en masse. It is concerned with 'dividuals', which are then 'governed automatically through databases and levels of access and exclusion' (Whitson, 2015: 343). Trust is no longer required to support these forms of governance. Instead, trust is set as the goal of the game, with the public not even as a competitor but as a resource to be exploited. Gamification objectifies populations, it datafies social power relations, and extracts legitimacy not from individuals but from dividuals and populations. To the powers that play the game, at the discursive level, gamification is the *manipulation of populations as data*.

To return to our examples, menstrual cycle tracking apps gamify at different levels. Without explicitly labelling predicted (or 'normal') cycle length as a score, the success rates of prediction based on average personal data create a relational score for the app over the individual: how well can it predict *my* cycle. Meanwhile adoption rates like downloads and monthly active user numbers become the wider game between competing apps. Amazon Alexa's voice of the dead shows gamification's ambivalence to human experience. Here it reflects a flattening of the experience of death to operationalize it as a tool for behaviour modification. Its effect on those who remain is reflected in the extractive logic that uses datafied human experience to improve trust measures amid attempts to expand usage into new areas.

## Consentification

If datafication and gamification refer to the manipulation of people (individuals and populations respectively) as data, how then is legitimacy extracted for these processes of governance, these logics of technology discourse? As we head towards trustification, we first look at the *extraction of legitimacy from individuals* within technology discourses.

This process is consentification. It will be familiar to anyone who has clicked 'accept all' on a website's cookie popup without trying to understand pages of impenetrable privacy policies. Within such contexts, consent is extracted through deceptive design patterns ('dark patterns') that trick users into giving consent by placing barriers to understanding and refusal, essentially denying the ability to mistrust. One-directional conditions of trust are extracted from the individual to a powerful actor in a way that makes consent itself function as a technology mediating the conditions of inclusion in technical and social relations or resources.

Zeynep Tufekci (2014) discusses how computational politics is about engineering publics rather than engineering systems (or, here, I would say it is about engineering the sociotechnical systems of power in which technologies, publics and discourses intertwine). This is a process of 'manufacturing consent' for 'obtaining legitimacy'. Though it happens at scale, often automated, the combination of (or negotiation between) individual rights and data politics/ economics means that consent operates through an iterative process of extracting legitimacies from supposedly informed and autonomous individuals.

This process emulates a contract and, as Elinor Carmi (2020) describes, contracts are where consent has traditionally been used. However, with datafied consent (and online platforms in particular) we see a shift from the static and specific nature of contracts towards expansive and elusive online agreements. As Carmi outlines, 'people have no way of engaging with and understanding what they actually consent to' and consent therefore becomes 'naturalising and normalising [...] a control mechanism' (160–1). Carmi (2021) furthers this problematization with the feminist critique that consent cannot be made more 'ethical'. It only strengthens and legitimizes the 'broken ecosystem' while placing the burden of responsibility on the user. This is what Becky Kazansky (2015) calls 'responsibilisation'. The inaccessibility of data contracts (whether we cannot find them or cannot understand their obstructive legalize) means that trust is both a requirement and output. Trust produces its own conditions as it is manufactured through the smoke and mirrors design of vastly asymmetric sociotechnical systems and power structures.

The function creep of consentification is also evident in technology discourses. This is seen in tools like the pixel and the tracking of individuals by major platforms even on third-party sites. Facebook, for example, holds a dataset about an individual as a Facebook user and a separate, though connected by default, dataset about that individual off Facebook – the latter even without ever having signed up for the platform. Consent within a user agreement is extrapolated within the discourse to encompass areas outside any such agreement. This consent, and the legitimacy for the entire third-party site tracking system that underpins many aspects of the online advertising industry, for example, is extracted unwillingly and unwittingly from users across the web. Even within the design of consent extraction systems, the dark patterns involved – which persist even after regulations such as the General Data

Protection Regulation (GDPR), which aimed at instilling meaningful consent as a sociolegal requirement (Nouwens et al, 2020) – rarely offer users any functional agency in the process.

Some platforms are even bold enough to perform this extraction openly, with companies like Consentify offering website cookie consent-as-a-service, leaning on regulators' acceptance of implied consent to shift consentification into the background of online activity. Within the infrastructures and dependencies of the web – think, for example, of the sheer scale of Google Analytics operating in the background of so many websites – the domination of major platforms and their design principles exploits trust for the transfer of agency across sociotechnical assemblages. Consent is itself datafied in the online ecosystem, extracted in the same manner as other forms of data, just another data point that is not only extracted but assumed. In Jathan Sadowski's conceptualization of data as capital (2019), consent would therefore stand in as social capital, a proxy to legitimize authority by manufacturing not trust but a representation of it within an asymmetric power relationship between individual user and platform.

Sara Ahmed writes that consent more generally turns 'strangers into subjects, those who in being included are also willing to consent to the terms of inclusion' (2012, 163). Consent is entwined with surveillance – just think of the minimum level of data collection required with cookies where even recording lack of consent requires either tracking or the constant repetition of consentifying processes. This builds on datafication and gamification as processes of subjectification and the violences they entail. A consent framework requires subjects to be known to a surveillant system whether consent is successfully extracted or not. These 'terms of inclusion' are picked up by Anna Lauren Hoffmann (2021), who highlights how such discourses increase vulnerability through data and discursive violence, all while trying to hide the extractive processes at work through the veneer of consent. This reminds us to keep power relations at the centre of our critique of trust, and to interrogate the way that consent is mobilized through discourses to legitimize processes of extraction, particularly when these processes are iterated and aggregated across individuals in ways that approach collective politics.

In our examples, individuals are consentified in menstrual cycle tracking apps as they are with many other data-hungry apps. The terms and conditions users are required to accept in order to use the app might conceal abusive data practices, but they also perform a discursive function in consenting to view one's own body in a certain way: as something to be tracked, computed and predicted. The self-tracking nature of these apps enforces consent through complicity. For Amazon's Alexa using the vocal patterns of deceased relatives, we see the hidden effects of consent even after death, and the loss of rights like the right to be forgotten that individuals suffer upon death. While Alexa becomes possessed by the dead, the individual themselves is dispossessed of their humanity.

## Trustification

My use of the term trustification here emerges from these concepts as a process of quantification, a logic of governance, and a legitimizing discourse. If consentification is the extraction of legitimacy from individuals, then trustification follows the same logic at the level of populations. It is the biopolitical legitimacy of institutions, of systems, and of public discourses, initiatives and technologies that, nevertheless, forward specific agendas. In doing so, just as consentification metricizes and operationalizes the agency and trust of individuals, trustification metricizes and operationalizes the agency and trust of peoples, communities and populations. Where consentification automates and anonymizes the process in relation to individuals, trustification functions more explicitly at the level of discourse.

Trustification legitimizes the use of technology to define the lives of peoples and populations, and the technologization of those lives into automatable units. It does this by converting trust into a monodirectional measure, denying the productive antagonism of trust and mistrust in political settings. Trust is assumed as an inevitable target.

Like other processes that collapse the qualitative into the quantitative under systemic agendas, a major part of trustification is this act of closing off an issue as something countable and solvable: the qualitative act of rejecting qualitative information. This has implications. Firstly, it deflects responsibility in seeking metricized legitimacy to account for public policies. Secondly, it erases wider lived experiences, existing inequalities and the spiralling issues created by those holding structural power.

Trustification is a process of legitimating these – often solutionist or economically-motivated – inequalities that exacerbate the injustices of mainstream technology discourses. This is the way that trustification emerges as both a process of quantification and a discourse that supports such quantification, both in service of extracting the conditions of trust from populations when relations of trust are absent or failing.

The foundations of trust, the reasons to trust and the reasons we deem someone or something trustworthy, form the basis of 'instruments of governance' (Nooteboom, 2002: 107). If trustification extracts legitimacy from populations, eliding the need to build trust, then it also operationalizes trust in its entirety as an instrument of governance. By metricizing trust at the level of populations, trustification performatively constructs trust as a prop for governance. It operates on the level of discourse, agenda-setting, and programmes of activity, rather than the individual contractual mechanisms that define consentification.

Though trustification may function within specific contexts, it is often tied in with the blurring of those contexts. It works through social function creep and the sliding of responsibility across the different (business, political, technical) actors and narratives that constitute the sociotechnical assemblages of power surrounding the use of technological objects in society. Trustification, operating at the level of populations, expands the responsibilization (Kazansky, 2015) of individuals into a shifting of responsibility onto entire populations. It becomes the public's responsibility to support technological discourses of progress, creating ensuing issues of inclusion and exclusion, of blame-shifting and scapegoating, and of the embedding of ideological hierarchies. The burden of trust is shifted onto the subjects of technology to encourage (or enforce) the performative constitution of trust.

Trustification sets trust as a target to 'get' out of a population, a resource to be gamed and extracted. But it also sets trust as a requirement for the population to perform. Technology would save us, repeats the discourse, if only people would trust it. And if you individually trust it, then it must be other people, other groups within society, who lack the trust to support the aims that technology discourses seek legitimacy to enact. This places the blame on society rather than those in power, dividing populations into those following the programme and those not fulfilling their duty to the narrative.

This furthers Hardin's complaint (2006) that discussions of trust tend to ignore trustworthiness, which powerful actors are no longer willing to prove or provide. Instead, it is the public that is meant to be trustworthy, and 'trusted voices' like community leaders are pressured to promote adoption. Like legal obligations, trustification tends to devolve trust to a minimum requirement by governments, corporations, research organizations and others holding power. It becomes the smallest measurable amount necessary to create a narrative of legitimacy. Meanwhile on a discursive level it denies the importance of justifiable and affective mistrust that might highlight historical and systemic injustices (of race or gender, for example) and call into question technological 'solutions' such as medical experimentation or surveillance. Trustification constitutes discourses that deny or hide the power asymmetries within sociotechnical assemblages while reinforcing those same asymmetries.

Returning to our examples, menstrual cycle tracking apps embody discourses of quantification of bodies, especially of marginalized bodies such as those of women, trans men and some non-binary folk. They use this to extract legitimacy for the application of generic technologies to specific, intimate areas of life, and the wider data ecosystems that exploit them. They also transform themselves into platforms to manage social interactions and support networks. While these tools can be used to spread information about menstrual health and other important issues, trust is played off against other forms of power, such as the Flo app (among others) making a public show of moving against the US government by enabling anonymous mode in the wake of the repeal of *Roe v Wade* as a way of protecting women's bodies against invasive laws against abortion and bodily autonomy while continuing the underlying data-sharing processes. This is again a very important tool, but also feeds into a 'trust us, not them' mentality that operationalizes trust in one social issue to support trust in a sprawling technical system with extractive data practices.

Amazon Alexa's use of deceased relatives shifts trust across networks of social relations. Families of the deceased are trustified into the Alexa system, and the initiative seeks to perform a proxy chain of trust for the userbase more widely. Legitimacy is extracted for the exploitation of data under the guise of emotional connection, to further the aim of trust and embedding technology within domestic life. Trust in deceased family members (or friends) becomes a proxy for trust in Amazon's aggressive data ecosystem, with terrifying potential for manipulation and abuse, while trust that the technology works proxies for trust in voice recognition and generation systems that legitimize these technologies in other areas, from access control discriminating based on accent or dialect to deep fakes that undermine trust in wider media and politics.

## Proxy variables

On a more practical level, trustification operates through constructing specific measures of trust. Because loci of trust remain always elusive across sociotechnical assemblages, these measures only operate as proxy variables. Just as consentification converts individual agency into the proxy variable of a set of binary accept/reject data points, trustification operates through proxy variables at the quantitative and discursive levels. In this sense it also echoes gamification in bypassing attitudes to focus on behaviours, specifically compliance. It attempts to stymie what Zuckerman (2021) envisions for mistrust as a driver and tool for transforming institutions and norms, by circumventing any discussion of frustrations or injustices and keeping the focus on metrics and outcomes.

Trustification does this by measuring trust only indirectly. It is certainly not about understanding the causes of trust. Trustification is measured always in terms of adoption rates, projected uptake, and other largely economic approaches to (for example) public health or other social concerns. Trust itself becomes not only a resource but a proxy variable (for social and political capital, for power) when it is operationalized in trustification. Again, trustification stands in for trust which stands in for trustworthiness, displacing the focus and burden of responsibility.

Glaeser et al (2000) found that surveying for political and generalized trust revealed trustworthy behaviour more than trusting behaviour; measuring trust (in society, in organizations) functionally acts as a proxy for measuring the trustworthiness of those surveyed; it is a check on populations not power. This echoes Hardin's (2006) focus on trustworthiness and the turning back of measuring trust not only as an obligation to trust but also an obligation and exploitation of trustworthiness onto those from whom trust is extracted. Similarly, Liu et al (2018) found that trust was a proxy measure for social capital – particularly at the higher end of both measures – and therefore viewed trust as a 'synthetic force' of cohesion from the interpersonal to the institutional.

Data about collections of individuals stands in for descriptions of populations as a whole, establishing an imaginary 'normal' or 'average' person against which everyone is measured. For example, a perception of trustworthiness is extrapolated and performed as a proxy for other perceived social values (like greater economic and social status, age, gender, race, and/or political conservatism). This becomes embedded in technologies like facial recognition in ways that perpetuate existing inequalities in the quantification of trustworthiness as a tool to be used against the marginalized.

Trustworthiness is performed as both a cause and effect of power, particularly when assessed at a distance. We often have to trust indirectly: individuals we do not know (strangers, particularly online, but also politicians or celebrities); institutions (governments, companies, organizations); and discourses (academia; the media; the tech industry; political ideologies). When trust is expected at a distance, shifted across relational assemblages, it is founded on norms, values and habits (Nooteboom, 2002: 65). This highlights the importance of discourse and the performative constitution of legitimacy for governance that is the focus of this book. Trust and control have been debated not as antitheses but as complements or even substitutes for one another, to greater or lesser extents (2002: 11). This is particularly notable in innovation, where trust and control are seen as mutually substitutable (Nooteboom, 2013: 107).

However, I would argue that innovation discourses also push for control where trust is lacking, and take the assumption that innovation is itself a valid end. The discursive performance of trust facilitates this proxy whereby legitimacy is forcibly extracted over publics, over regulation, over competitors. This is the eliding process of trustification, the slippage from trust to control, the avoidance of scrutiny. Under this performance, trust is constituted as a proxy variable for power, and this power stems from control rather than democratically bestowed authority.

The decline of historical markers (of class, education) has been seen as an, often positive, contributor in the decline of traditional proxies of trust attributed to factors like more educated and critical publics (Norris, 1999), and the role of trust as a heuristic has been discussed with increasing importance in political discourses in the wake of this shift (Goubin and Kumlin, 2022). But this process has led to new proxies, particularly within technological discourses and often through forms of monitoring and control (datafication, gamification) to make behaviour more predictable (Boswell, 2018: 40). The markers may be less explicit, but the power asymmetries and injustices are just as entrenched.

These forms of control as proxies for trust are mediated by norms that reinforce measurement, rules and automated responses, diminishing not just the requirement but also the possibility for informal, interpersonal models of trust. This leaves loopholes in the sociotechnical assemblages that enable the exertion of power and enables measurement to descend into a spiral of extraction, a drive towards authoritarianism lying beneath the liberal and individualistic narratives of technology discourses. Again, this comes down to manipulation and compliance.

Trust, within trustification, is the ability to manipulate, to control. It is the legitimation of sensory power (Isin and Ruppert, 2020), and the imposition or assumption that the governed will accept such power. It therefore feeds into issues of normalization and controlling social narratives (especially around technology discourses and what is constituted as permissible or desirable under labels such as 'innovation' or 'for good'). Fundamentally, trustification operates as a constant slippage of measures to deflect underlying questions about power, and the ability to hide agendas in setting the discourse of any given sociotechnical context or of technology in society more widely.

This chapter has outlined the emergence, function and role of trustification in legitimizing technology. In the following chapters the discussion focuses on different logics of power in which the process of trustification operates; in turn, I examine state, corporate, research and media forms of trustification that perpetuate each context's own agendas in performing the conditions of trust in technology.

# THREE

# State: Measuring Authority

The previous chapter outlined the ways that trustification emerges. In this chapter, I apply this conceptual framework to state logics of power, and the different ways that states have trustified their populations throughout history and today. From projects of measuring through to large-scale surveillance and manipulation, states have unevenly extracted legitimacy through the use of exploitative technologies.

Trustification turns trust into a technology of power. As with many technologies, and many forms of power, trustification is not a brand-new phenomenon. Despite the representation of constant newness that technology discourses use to maintain the legitimacy of the ideology of progress, technologies tend to reflect, reify and reinforce existing social values, inequalities and injustices. So too can the underlying aims of trustification be traced backwards in time, particularly as embodied in the state.

The history of representative democracy is a precursor of trustification. This 'indirect' form of governance rests on measures of authority. Power slides and aggregates across social assemblages. Narratives of votes, representatives and majorities act as measures for the legitimacy of a political manifesto. Confidence and trust produce a mandate with which to exercise authority, played out with higher levels enabling more extreme policies. Games between those in or seeking power treat populations as a resource from which to extract legitimacy for that power.

It is useful to consider the mechanisms and moments in which state or representative authority breaks down, the points where the extraction of legitimacy becomes visible. Election monitors meant to prevent fraud are designed as a mechanism to support (or reject) the legitimacy of the outcome. Australia's representative democracy enforces participation as a means of extracting legitimacy for the wider system: abstention is still registered as a mark of consent; and the historical exclusion of indigenous peoples categorizes who does and does not count as the population from whom legitimacy is required. Autocratic regimes employ unfree and unfair elections to force consent from their population, falsifying measures to maintain legitimacy. Migrant populations suffer bureaucratic and now technocratic counting to 'count' for aid (Metcalfe and Dencik, 2019). This extracts the conditions of trust out of sheer necessity. Among those embedded within the games of power, a vote of (no) confidence quantifies even mistrust, shifting the role of trust to be operationalized as a power grab under a metricized veil of democracy between those playing the game.

These extractions of political trust are counterbalanced by ongoing levels of mistrust among populations. There is a politics that runs beneath the mistrust in the principles, institutions and practices of law (Tyler, 1998). As Zuckerman reiterates, 'mistrust is unevenly distributed' (2021: 20), and this carries a line from historical injustice through to today's operationalization of trust by institutions. The historical component identified in mistrust echoes its colonial roots and the ongoing effects of slavery (Nunn and Wantchekon, 2011). This draws parallels with the rise of mistrust within specific neighbourhoods where social injustices and the failures of institutions are most apparent, and where a sense of powerlessness from systemic issues amplifies feelings of mistrust (Ross and Jang, 2000; Ross et al., 2001).

The injustices suffered by sections of populations racialized and gendered as minorities (and the escalation of such injustices for people at the intersection of multiple forms of discrimination) continue despite increased rights and other legal protections. Legal recognition ('counting') for trans people, for example, perpetuates both administrative and thereby physical violence (Spade, 2015). I suggest that the increase in trustification comes about within the datafied narratives of society as the response of power to the combination of escalating mistrust by minoritized sections of populations and the appearance of wide-scale mistrust in otherwise privileged sectors of society. Trustification is an attempt to close off mistrust, to close off alternative discourses and alternative power structures by continuing to extract legitimacy through performing the conditions of trust.

## Trust, technology and the state

Twenty years ago, surveillance scholar Oscar H. Gandy (2021 [1993]) described the phenomenon of manipulating people and populations as data as 'the panoptic sort'. He identified that while 'the question of trust is a key component in the formation of consciousness and the demand for state action' (2021: 47), the panoptic sort 'does not engender trust and a sense of community. Quite the opposite is the result' (2021: 260). States have contributed to degrading trust by replicating within their own discourses, agendas and priorities the technological and datafied narratives that are already imposed on populations. For example, Boswell (2018: 38) discusses in detail the UK government's adoption of a target-driven approach to immigration in the 1990s to 2000s as a failure of policy, one which we can identify as another step on the path to technological solutionism. This trend feeds into (often racialized) datafied policy such as the embedding of surveillance within border controls, or in the personalized politics that traces individuals not as individuals but as datapoints within a population.

Engin Isin and Evelyn Ruppert (2020) add to Foucault's biopolitical path through sovereign, disciplinary and regulatory power with a fourth category of 'sensory power'. A sensory logic of power uses a strategy of performance, using modulation and constructing 'clusters' as assemblages or categories defined after the fact. These clusters are relational, multiple, fluid, visualized and live: the dashboard as governance. In terms of trust, it is the ever changing and shifting of proxy variables that trustification is willing to adopt. Trustification cares not which metric is used, only that it can be used to extract legitimacy. The boundaries of populations can be manipulated, gamed, to justify these ends after the fact, performing and constituting legitimacy out of continual pulses of information.

Isin and Ruppert show how sensory power was made visible during the COVID-19 pandemic, in which various shifting metrics were used to manipulate people and populations, all the while seeking legitimacy for solutionist narratives created out of the competing agendas of governments, corporations and research institutions. Trust became a proxy for compliance in public health measures during the pandemic (Van Bavel et al, 2020). The decades-long decline in trust rooted in failed institutional performance (Newton and Norris, 2000) became a theme to justify failings or barriers encountered in public health policy and measures during the COVID-19 pandemic, with a 'trust deficit' being blamed for harming government and technology responses.

Extracting trust in measures, and compliance with the policies they support, requires the same balance of demonstrating efficacy and accountability required for wider political trust. If we acknowledge that trust 'shapes, and is shaped by, policy responses in complex ways' (Devine et al, 2020: 9), then the performative constitution of trust as measurement emerges in relation to wider political discourses and sociopolitical assemblages. During the pandemic, different contexts around the world emphasized not just trust but whether trust is a choice (Abbas and Michael, 2020), the 'conditions' rather than measures of trust (Greenleaf and Kemp, 2020), the role of legitimacy, information and transparency in government use of social media in building or damaging trust (Song and

Lee, 2016; Limaye et al, 2020), and the way communities stepped in to replace low trust in governments (Hartley and Jarvis, 2020). Across these different contexts and issues, trust is moved across complex sociotechnical assemblages to fulfil government policy aims.

The compulsion to metricize trust has become increasingly dominant in global narratives that seek to compare states and their authority. As the OECD writes, 'the relevance of measures of trust is not in doubt [...] to assessing the well-being of societies, to measuring social capital, and to understanding the drivers of other social and economic outcomes. The accuracy of trust measures is less clear' (OECD, 2017: 16). This framing raises issues of accuracy without challenging the conceptual basis of measuring trust or the sliding of interpersonal trust within populations onto trust and power of institutions and states. It thereby gamifies populations as a resource for states to use, competing to see who can extract more legitimacy.

The ends of authority also support the means, incorporating the drive towards measurement, quantification and datafication within how authority is conceptualized. Trustification legitimizes the authority of individual governments, but it also legitimizes the discursive games being played between states. This game is also turned inwards, playing different segments of populations against one another, to distract from building political action from collective mistrust in authority.

## Trust us, not them

If trust is relational, it echoes existing comparisons and inequalities of power relations. Simone Browne's (2015) detailed discussion of state surveillance of Black communities by US law enforcement shows the racialized histories of tracking populations. This demonstrates a further inequality and displacement seen in trustification. Data is extracted from minoritized groups so that legitimacy can be extracted from majoritized groups. People are not counted equally.

Being counted can be quite different from being someone who counts.

James Scott (1998) showed how the modern state is based on the ability to measure, name and act at the level of populations. This leads to seeing everything according to standardized patterns and bureaucratic categories. Building on this, Dean Spade (2015) outlined the concept of administrative violence, particularly in the experiences of trans people of colour. The assertion of state power through trustification and establishing unjust norms of who *is counted* versus who *counts* demonstrates this long-running problem of administrative violence. The apparent neutrality of administrative systems enacts power in the construction of norms while denying that process of norm construction.

Administrative processes are performative in the way they constitute oppressive discourses of homegenization and standardization to constitute populations. As Spade writes: 'power relations impact how we know ourselves as subjects through these systems of meaning and control' (Spade, 2015). These systems are often the places where racism, ableism, sexism, transphobia, homophobia and other systemic categories of oppression are constituted. Yet they are performed as if they are neutral. Trust is operationalized to separate the accepted population from those who do not count.

The division of an us and them, of citizens and non-citizens, of data and people, uses standardized measures and categories to delegitimize the claims of subjected groups to 'counting' (Beaugrand, 2011: 229), and offers only these data-driven options of conformity, of quantified roles to perform. Trust is extracted from those who do count, who are counted, in opposition to external forces outside the state – whether that is those subjected to state violence or other forms of power. Trans folk seek recognition essential for healthcare; refugees must be counted and tracked to qualify for even basic resources. All the while, politicians and media mobilize narratives that blame these groups for policy failings and legitimize violence against them, dis-counting them from majority society. For the privileged, being counted counts. For those who are marginalized, they are merely kept count of.

These externalities of state trust can be seen within technology discourses across the UK's expanding range of regulations around different facets of digitalized society. There are tensions across data protection and online safety regulation, alongside related issues in advertising standards, children's rights and competition (to name but a few). These conflicts highlight an antagonistic approach to policy making that seeks, at base, to assert state power over wider areas.

The overarching discourse still maintains a datafication narrative. The UK government developed a National Data Strategy from 2019 to 2021 that pushed for increasing collection and use of data in the public sector. The strategy also promotes increased exploitation of that data in the private sector, and half the initial engagement roundtables that informed the strategy had an industry focus. This highlights the porous nature of the state and the tightly intertwined systems of government and corporate power and actors.

Similar data narratives were responsible for limitations in Brazil's data protection legislation. The assumption of invasive data practices was pushed as 'officials argued that sharing data is a recurring and essential practice for the daily operation of the public administration' (Reia and Cruz, 2021: 230). But against this technological expansionist frame lies a tension with other forms of power.

Data protection and online safety regulations are key examples of these tensions. While the UK's Information Commissioner's Office (ICO) asserts a record of fines for data protection breaches, they are largely focused on companies for whom data is not the primary business. The ICO has been far less willing to directly tackle issues within the data ecosystem, such as major platforms or data brokers, despite releasing multiple reports on the systemic harms to individuals and society by data and platforms ecosystems. This is perhaps why the development of online safety regulation has been touted as a world-leading step to deal with these problems, making notable claims of accountability to be imposed on these big platforms.

The game continues to be played between different forms of power, with populations (like the data about them) treated as a resource to be extracted and exploited. But within this is an assertion of state authority over the Internet, leaving definitions continually open to debate. Various revisions even more worryingly shifted to less scrutinized policy more directly guided by a single government minister – highlighting the imposition of state power over online discourse. By leveraging legitimate concerns like child safety within a populist sensationalized framing, the government sought trust in its ability to protect its citizens online by being the first to impose such regulation. But by pushing towards a censorship-heavy model and proxying harms onto important mechanisms like anonymity, the same old administrative violence is performed again and again. The harms left open to trans communities, for example, are built into the regulation by design, translating culture war discourse and the categorization and subjection of certain groups to extract trust from the 'accepted' citizenship in the state's role of security in online environments.

Similar games of power and discourse, with harmful consequences for marginalized populations, are discussed in Sarah Brayne's analysis of how policing leverages data and surveillance to widen the scope of criminal justice (2020). Trust and sensory power emerge again under the technologized narrative of safety to expand sociotechnical assemblages, performing prediction narratives as self-fulfilling and self-justifying truisms. The division of those whose trust counts and those from whom trust is simply extracted reduces both 'us' and 'them' as resources from which legitimacy can be extracted for the overarching narratives of authority. These narratives constitute the datafied categorization of accepted citizens and subjected others. For example, if you only look for crime in certain areas you will only find crime in those areas, replaying the 'redlining' of minoritized neighbourhoods through metrics that erase context, alternative knowledges and systemic injustices. As the state trustifies its expansion through increasing datafication, those boundaries become more administratively violent.

Building a sense of us and them gets carried into situations like states of emergency that may have no direct other. Elizabeth Ellcessor describes how in the mediation of crisis or emergency, 'technologies further obscure the processes of decision-making, not to mention the political economy of funding, data and operations' and to some extent 'the production of an emergency by an alert is a matter of trust' (2022: 93). Such an alert is performative. It creates the emergency by labelling it, in ways that normalize and obscure inequalities in the burdens, harms and extractions the emergency is used to legitimize. Trustification conceals these different experiences of emergency with reductionist metrics that attempt to constitute a sense of unity to assert an apparent duty of the population as a whole.

The increasing reliance of states on technological solutions – especially during emergencies such as the COVID-19 pandemic – creates an 'inextricable link' (Bodó and Janssen, 2021) between a state's perceived trustworthiness and the (often lack of) trustworthiness of the private companies providing such services. We can reiterate the performative nature of these processes here. Constituting an emergency (and the legitimation of additional authority, powers and policies with reduced scrutiny that a state of emergency entails) requires legitimacy often based on trust. But where to place that trust becomes obscured within national(ist) and/or humanitarian state discourses. The burden of trust becomes enforced as a duty within such discursive extraction of legitimacy. The 'them' that trustworthiness is constructed in relation to becomes formed of those members of the population who do not comply, who do not fit, again asserting a division of counting to decide who counts. These inequalities are reinforced not only in games of discursive power, but in the concrete policy outcomes that such power legitimizes.

## Not so smart policy

A famous sociological maxim states that 'not everything that can be counted counts, and not everything that counts can be counted' (Cameron, 1963: 13). The issue of counting is particularly pertinent within technology policies around AI, the push towards data-driven society and the increasing 'smart'ness of our lived environments. For example, I have already mentioned the EU's High Level Expert Group on AI, which spent two years developing substantial policy recommendations focused on trustworthiness in AI. Many of the members of the group have made significant contributions to ethical, legal and social aspects of technology. But this framing of trust has led to several problematic outputs, including a deliverable specifically aimed at helping organizations measure their own trustworthiness. This is trustification not only of the supranational state mechanism but the passing of that legitimacy from such power down and across the sociotechnical assemblages within which AI discourses operate.

These organizations push narratives and tools that legitimize measurable proxies for trust, fairness and other social relations. Measuring principles like trust or fairness assumes thresholds and acceptable levels for legitimizing wider programmes of technology and policy. This removes the important need for space in which to highlight the option to not develop, design or deploy certain types of AI or AI within certain contexts. Instead, we must emphasize respect for the right of refusal or resistance. We need to acknowledge that trust must be built out of a relation with critical mistrust, as opposed to operationalizing trust in technological artefacts as a proxy for sociotechnical relations with massive power differentials.

These asymmetries in power relations can be seen between individual states and global state frameworks. For example, Reia and Cruz (2021) highlight the critical contribution of Brazilian experts to the development of the UN's New Urban Agenda. They problematize the continued 'commitment to adopt' smart cities embedded within the overarching aims of the agreement. This agenda-setting work was adopted at the UN Habitat3 event in 2016 and endorsed by the UN General Assembly later that year. The agenda established a clear discourse that assumes all cities are and should be heading towards policies committed to deploying 'smart' systems (and the increased surveillance, data and algorithmic practices this entails). The agenda emphasizes 'well-planned and well-managed' cities (Habitat3, 2016), echoing data narratives backed up by trustification, a performative politics that constitutes adoption as a need by saying it is a need.

Reia and Cruz show the absence of key questions in policy discussions of smart cities: 'privacy, data governance, and the right to the city'. They emphasize the need to 'understand the power relations in the intersection of infrastructure (materialities), policy, and politics' (2021: 229). Different facets of the sociotechnical assemblage of any given technology (and its discourse) introduce different agendas playing games of power with populations supposed to trust not only in the individual powers but also the game itself.

And yet, alongside the shifting of burdens of trust onto populations, trustworthiness is also shifted as a burden onto populations (as we saw in administrative violence discussed above). But the ability to mistrust is in turn shifted onto the state as an additional power, legitimizing increased surveillance and data-hungry narratives that enable greater monitoring of populations. And these drives often perform racialized narratives and other injustices. As Achilles Mbembe (2019) describes, 'the dream of perfect security [...] requires not only complete systematic surveillance but also a policy of cleansing'.

Complete technological solutionism is only ever complete or a solution for some.

For example, facial recognition is increasingly used directly by police but also embedded within the 'smart'ness of cities and devices. These technologies convert mistrust of populations by the state into a technical artefact. The systems treat everyone as part of a permanent police line-up, while avoiding visibility, responsibility and trustworthiness of the systems and their operators. They are designed to see but not be seen (through). They continue to be used despite known flaws and biases along racial and gender lines (Buolamwini and Gebru, 2018). If you try to hide from these untrustworthy systems' attempts to judge your trustworthiness, you are confirming your (un)trustworthy status to the wider social system. This not only raises issues of cultural and racial discrimination in the Global North against covering a face (whether with a hood or a veil), but also the response of power. For example, London's Met Police harassed and fined individuals who sought to cover their face while walking through an area under 'voluntary' trial of live facial recognition.

Facial recognition embodies the quantification of people as a new form of 'cleansing' that removes the messy complexity of human experience, expression and relations, cleaning society by cleaning data according to technological and colonial logics. It demonstrates the entrenching of biopolitical and sensory power within the material, legal, political and discursive fabric of our lives. It deflects trust across the sociotechnical assemblage and into different power levels of technological discourse. This opens further games between states that reify international power relations based on historical global inequalities.

## Colonizing legislation

The narratives that shape the role of technology in society continue to operate according to colonial logics. This builds on the economic meaning of trustification, with its monopolizing effect applied to the way different countries were brought into EU narratives of cloud computing and the economic agendas that dominate policy making with 'utilitarian computer science' narratives applied to 'administering information webs' (Lapuste, 2015). More outward-facing processes can be seen in the influence of NIST from US policy concerning scientific and technological research and industry into setting worldwide discourses around standards, measurements and protocols.

Whether units of measurement or cryptographic protocols, this influence exerts a dominance over not just specific technologies but the frameworks within which they are developed. It sets the scope of future development by shaping the way they are understood and described. By taking a leading role in establishing internal standards, combined with the dominance of the US in the tech industry and research, mechanisms like NIST establish their legitimacy as the defining discourse for the assumptions and practices of technology. This can be seen through, for example, the continued weighting of higher-rated conferences in AI or cyber security towards US-based researchers.

In the realm of technology policy and legislation, however, Europe remains a dominant locus of colonial discursive power. By combining the influence of these historical powers, and through proactively seeking to establish regulation first, the EU uses its vast administrative power to enact discursive power over technology laws worldwide. The entire framing of the huge number of documents the EU produces among its different divisions and expert groups constitutes the legitimacy of the organization by appropriating academic and industry influence as well (seen in, for example, the High-Level Expert Groups), manufacturing an ecosystem of trust.

The titles of these documents clearly demonstrate the values and narratives embedded in these proposals, such as the European Commission's (2020) White Paper on AI entitled 'A European approach to excellence and trust'. This aligns a specific set of values with trust not only in technological systems but in the political dimensions of sociotechnical assemblages. We can view them as an ongoing attempt to colonize AI legislation worldwide with European values. This includes specific conceptions (and limitations) of individualized rights, informed citizens, and what constitutes public good within an ecosystem of trust. The 'Brussels effect' of how EU policies spread beyond Europe is clearly seen in technology policy, and its spread is not simply a matter of accident or leading by example.

Laws are a key export of the EU, trading in influence and manufacturing the legitimacy of the EU to speak not just for Europe but for the world. The GDPR is a prime example of this. Some African nations such as Benin, Uganda, Egypt, Kenya and Nigeria have followed GDPR models for data protection legislation – particularly on extra-territorial aspects. The Republic of Cabo Verde previously modelled legislation on the older Copyright Directive and updated it in 2021 to follow EU regulations.

This could be seen as a neo-colonial expansion of setting regulatory narratives, values and priorities by the EU. It could also be seen as a safer bet for other nations to follow suit knowing they have an established set of laws from a combination of powerful nations to stand up and protect its citizens against exploitation. This exploitation comes from third-party countries like the US and now UK who push narratives of deregulation, playing games of power with the EU to impose a different set of values and agendas that support exploitation by multinational (but most often US-based) tech giants who make use of a favourable lack of regulation. Both the spread of EU-based legislation and the global push for US corporate narratives also demonstrate the cascade of trustification to support adoption and ratification internally within other countries considering legislation. Narratives of national interest against global threats are combined with either the established power of European legislators as a secure model to follow to achieve firm regulation, or the US-corporate image of economic power and prosperity that accompanies a deregulation model.

Trustification not only acts upon populations but also throughout the levels of institutionalized power. Politicians and policy makers down the ranks and across parties or ideologies are also trustified, converted into a different form of vote. This is gamification of legitimacy for policy making within states and globally, carrying with it a sliding of power and agenda setting towards Europe.

The active influence by the EU on global policy discourses can be seen in the Harmonization of ICT Policies in Sub-Saharan Africa (HIPSSA). This initiative, launched in 2008, has produced subcontinent-spanning guidelines. It functions as a looser version of EU policy, like the 2013 SADC Model Law for Data Protection that sets the discursive priorities and norms for multiple countries. The project was funded by the European Development Fund, which exerts a financial metric of legitimacy on the process, and the model was published by the UN's International Technology Unit, adding further dimensions of global dominance over local and regional policy.

While optimists might highlight the facilitation of agency, colonial narratives are evident. The framing of the process echoes what Mbembe (2019) describes in relation to extra-legal power: 'in the imaginary and practice of European politics, the colony represents a site in which sovereignty fundamentally consists in exercising a power outside the law'. The externality to the law here is not physical violence but colonial administrative violence between states, the discursive violence that dominates the production of national laws and self-sovereignty under European values proxied through European laws and models. This logic spreads more widely in narratives of data sharing in the African continent, where deficit narratives imposed by Europe and North America – the perception of access and skills – frame barriers to adoption (barriers experienced by colonial stakeholders seeking to exploit African nations for data extraction) as problems of those the data is about (Abebe et al, 2021).

Just as trustification places the burden of trust onto populations, these colonial narratives reflect the shifting of expectation and responsibility onto subjected populations in order to extract legitimacy for the extraction of resources and capital (here also/as data). These narratives are embedded in the export of European policy and North American standards. The violence of performed neutrality emerges once again in global standards that are anything but value neutral, particularly when they are shifted onto globalized 'solutions' without contextual specificity. This is exacerbated by the ongoing push by many colonizer states towards extractive data economies and the support of economic interests and narratives over their own populations and others.

## Privatization by any other name

Louise Amoore (2013) details in post-9/11 state responses to the possibility of major threats a shift in focus towards 'low probability, high consequence' scenarios. War gamification abounds in the context of these logics, but also a significant increase in commercial opportunities for state security contracts. Even within intelligence, an area once kept separately by the state, there has been movements towards privatization. France, for example, 'entrusts' surveillance data to private companies for geolocation analysis (Bigo and Bonelli, 2019: 115), while the security market is ripe for the invasive, harmful and legally dubious tools and services offered by surveillance companies like Palantir and NSO who are surrounded by ethical and legal questions. Within the embedding of surveillance in cities – data collection everywhere – policy makers and researchers are often 'failing to pay sufficient attention to the increased corporatisation of the municipal governance entailed by most "smart city" proposals' (Reia and Cruz, 2021; 229). The different speeds and timings between sectors (advocacy, policy, legislation, industry) create opportunities for the private sector to fill the gaps.

Beyond direct contracts for tech companies being embedded within more and more areas of policy, tech policy itself has been increasingly privatized. Lobbying has become a huge area of focus for major corporations. Lobbyists function as proxy actors in the shifting and obfuscating of sociotechnical assemblages. Moves like mass lobbyist hires, targeted lobbying campaigns, and draining legal expertise to remove opposition form a set of policy dark patterns performed by major tech corporations. Huge amounts of funding are poured into these endeavours. For example, Cini and Czulno (2022) assessed 4,342 lobbying meetings with the European Commission between 2017 and 2021. They found the meetings to be dominated by big corporations, particularly big tech, represented by both the number of meetings and the level of access to senior policy makers.

The 'contours' of this lobbying represent 'a rare, visible instance of the interlocking relations between media elites and the state' (Popiel, 2018: 567). This shapes the scope of possibility within sociotechnical assemblages that surround technology and policy, in which public issues 'become "economized," gaining legitimacy only in the context of technological and economic growth' (2018: 580). Tech giants like Google (Riekeles, 2022) and Uber (Guardian, 2022) are heavily implicated in leveraging politicians individually and collectively to further their aims and remove barriers, converting economic power into a proxy for trust. In the context of mass corporate lobbying of technology policy, the discourses surrounding technology become fixed within corporate logics of solutionism, technology for technology's sake, a desire for innovation.

Rather than measuring authority through the proxy metrics of trust, state authority could be assessed according to the balance of public or corporate influence over policy. Otherwise, policy, and by extension the state, becomes merely a source of legitimacy for corporate techno-solutionist narratives even as the state seeks to trustify its population to legitimize surveillance and data practices to perform its own authority. The multidimensional discursive games of trust extraction continue.

# FOUR

# Corporate: Managing Risk

Here, I build on the previous chapter's discussion of trustification under state logics of power by focusing on corporate power. The companies that design and manage technologies wield inordinate power, with the ability to embed certain values and priorities. This 'governance-by-design' (Mulligan and Bamberger, 2018) risks bypassing state and public forms of governance through the ways technologies are developed to shape increasing areas of everyday life. In order to justify this power, and to keep it in the hands of tech companies, trustification is used in ways different from state logics, instead based in corporate narratives and logics of power.

Corporations engage in games of legitimacy against states and societies. Companies and states perform certain roles in relation to one another and constitute their power within such relations: as regulator; as provider of essential infrastructure or services; as investor; as problem to be dealt with or avoided. To accomplish this, each must learn to see (and speak) as the other. As Félix Tréguer (2019: 148–9) has identified, there are different constraints affecting a company's focus on resistance or cooperation with state surveillance, particularly following the Snowden mass surveillance leaks. These constraints include high concern for user trust and competition weighting more towards resistance, while holding (or seeking) government contracts weights more towards cooperation.

The games come down, in part, to where a corporation seeks to extract legitimacy, and managing potential risks. For example, pursuing a law-enforcement contract for facial recognition algorithms may make Google feel more inclined to share people's search data with that same agency. The straight extraction of data from individuals undermines the idea of any kind of transaction between corporations and their users, consumers or workers (Lyon, 2019: 67). Instead, data is exchanged between corporations (and states): games played at levels of power far beyond the populations the data is about.

The imposition of corporate discourses over several decades has embedded economic risk-based metrics into state thinking. The state also sees like big tech; the audience to corporate performances turns into a performer of those same discourses. For example, state capitalism 'now appears to deploy its authority less to consolidate its political rule than to reduce its market risk' (Gonzaga, 2021: 448). Neoliberal policies aimed at supporting (big) business have also performed those same risk-based priorities within state mechanisms, in turn making techniques such as lobbying more entrenched and/or effective.

The economic extraction of legitimacy through quantification has a long history. Many technologies are deeply historical. They can be rooted in long histories of development and discourse, and they can also function by replicating history. Data, algorithms, machine learning, surveillance all operate by measuring and reproducing the past to manage risk in the present and future. This function of technology can be traced to the conjoined development of colonialism and capitalism.

Today, unjust systems of measurement such as redlining have been converted into digital access (Gilliard and Culik, 2016), whether it is censorship within school web browsers or differences in Internet speeds between neighbourhoods. Canay Özden-Schilling (2016) charts the evolution of physical infrastructure such as electrical grids. This has been through the long-running process of partial deregulation, 'a project in reorganising information' (2016: 69). It has led into all manner of industries becoming informational services managing public services (and labour) as data according to market principles and computational logics. The social function creep of companies and their data logics has further embedded racialized (and other) inequalities in a false sense of objectivity, measurability, predictability and causal variables. These ongoing injustices form the basis for data not only as capital but as a discourse of capital.

## High scores

The history of accounting is a history of keeping score, of metricizing value and risk. Companies and technology discourses are entwined in attempts to predict risk. For example, credit scores such as FICO in the US since the Second World War have operated as 'calculative risk management technologies' (Poon, 2009: 268). These systems, often named for the private companies who manage them, escalated historical inequalities to create further avenues for abuse. They concealed the shifting of agreements across financial assemblages of securities and servicers. Credit scores were closely linked to the 2008 financial crisis. In the increasing interconnectedness of financial and other services, 'the rapid diffusion of credit scores [acts] as an index of trustworthiness' (Gandy, 2010: 33), pushing beyond their initial scope (often supported by permissive regulation) to shape more and more of people's lives.

Trust in these sociotechnical assemblages is expected (required) to flow upwards from individuals through technical objects (scoring systems made up of vast datasets and prediction algorithms) to reinforce the authority of corporations and corporate logics to justify these approaches to managing populations. It is this discourse layer that trustifies the legitimacy of the entire financial system despite the vast exploitation and the sliding of risk onto those already suffering the harms of economic and technological systems.

Eve Chiapello (2007) discusses the role of accounting not so much as the origin of capitalism but as the origin of the *concept* of capitalism. Keeping track of circulating capital allowed it to be articulated, and 'the concept of capitalism is indissociable from a representation of economic life shaped by an accounting outlook' (2007: 263). Accounting is the counting and monitoring of people and things understood as assets or resources. It performs the economic system and its sociopolitical implications as measurement and standardization that construct certain knowledges and politics.

Marion Fourcade (2011) problematizes the 'measuring rod of money' in relation to creating supposedly rational metrics for the value of intangible entities such as nature. There are contextual specificities of what that means to those extracting value versus those feelings the harms of extraction. As Dan Bouk details (2015), risk as a quantification of people's lives long predates the age of big data, with past measurements of individuals or groups constituting their futures as mere statistics.

Even more harmful discourses of colonial power surround the documenting, taxing and insuring of slaves, for example in eighteenth- and nineteenth-century Brazil (Rodrigues et al, 2015), managing risks of people not only as data but as property. Trust in a different form can be seen in Trustee Georgia, which gave way to both slavery and land tenure in the proto gamification of territory and populations as resources in the absence of 'trust'. During the slave trade, insurance operated as exoneration by describing people in terms of 'collateral damage' and 'unavoidable losses', but this framing of 'incidental death' belies a narrative where 'life has no normative value [...] the population is, in effect, seen as already dead' (Hartman, 2006: 31). Even after the slave trade to the US was outlawed in 1808, but before slavery itself was abolished, insurance took on a heightened roll to protect the increasing value of people as 'assets' (Ralph, 2018), the necropolitics of slavery as 'expulsion from humanity altogether' (Mbembe, 2019: 21).

We can contrast this to Black Founders in the US establishing institutions like insurance groups as safe spaces from racialized injustice. They did this by creating alternative power systems and social relations linked to 'critical intellectual agency' rather than logics of monetization (King, 2014). But in the power of colonial and racist finance assemblages, dominant narratives persisted. Mechanisms like redlining marked racialized communities for exclusion from access to finance. These discourses and social technologies of oppression datafied populations into a resource to be managed according to risk within the games between financial powers. This history, based on categorizations of people as human, not-quite-human and non-human (Weheliye, 2014), is one reason corporate data-as-property narratives provide little to tackle systemic injustice today.

There is an underlying conflict between the market priorities of the growth data economy and issues of social responsibility. This stems from shifting the burden of risk onto data subjects in the personalization of risks and emphasizing personal responsibility for decisions made by corporations (or governments or academia) through technological proxies (König, 2017: 2–4). The attitude to risk and measurement operates as trustification, extracting legitimacy for the sociotechnical system by measuring risk as a proxy variable for trust. For example, if you are low risk, you are trusted by the system, but the trust in the system making this judgement is implicit and extractive.

Risk comes down to power, displacing harms of risk onto others while maintaining the benefits. Corporate and technology discourses enable this. Risk therefore functions as an inverted proxy for trust: if risk can be accounted for, displaced through measurement and countermeasures, then the need for trust, it is assumed, reduces. But these problematic assumptions about 'predictable and trustworthy' technologies and policies in turn become embedded in public decisions in ways that benefit companies. For example, in March 2022 European Commission President Ursula von der Leyen announced a new EU/US data agreement that would be applicable to companies who move data between legislative areas, creating a legitimizing backdoor under the promise of trust and risk management.

The measures for this risk–trust relation are often displaced onto the performance of a given technology (or policy, initiative, or process). The ability of technology to function as expected – or at least according to levels determined by the interwoven networks of its manufacturers and regulators – stands in for trust. In this way, performance is performative; it constitutes trust in the sociotechnical assemblage through measures of a technical object. This can be seen in technology benchmarks and in policy. For example, the levels attributed to self-driving cars are aligned with the levels of risk – seen through legal liability – that their manufacturers are willing and allowed to take. These conceptions of risk often fall back onto issues of insurance (Shannon et al, 2021), a problem long associated with the automotive industry's cost-based analysis that balanced, for example, payouts against recalling flawed designs. There is a necropolitical power to these financial systems that continue to assign monetary values to human lives.

These logics turn back onto trust systems themselves, extending historical credit scores into wider social scores. The same mechanisms of the much-demonized state-led social scoring system operated in China have long been in operation behind the scenes across the capitalist world. This is often enacted by states, like the points-based immigration systems used in various countries like the UK. But they are often designed and owned according to corporate power. For example, Trust Science was granted a patent in 2019 for 'Calculating a Trust Score' that was notable for its aggressively inclusive range of data types to be used as well as its claims of general application.

This represents only the latest iteration in a long line of trust scores. IBM, Amazon, Google, Microsoft, Facebook, Cisco and Intel collectively own hundreds of 'trust score'-related patents, going back at least to the early 2000s. Other examples go back much further, the companies and patents being bought and sold as part of the ongoing corporate games of influence through intellectual property and market dominance. The idea of trust as a computable concept harks back to decentralization and distributed computing, intertwined with the history of AI development, in turn leaning back on mathematical and economic models of trust and legitimacy.

## Economic narratives and power

Money itself often functions as a proxy for trust. 'Money talks' as the saying goes. Like trust, money acts as a proxy for the ability to get things done, a marker of potential to complete a proposed agenda, with additional resources building up a greater bank of this faux trust not only to enact an agreed agenda but also in turn to set those agendas. Corporations, particularly technology companies with vast amounts of wealth or investment behind them, leverage this power to set agendas around the embedding of their technologies into areas of society, and shape policy, practices and wider discourses.

Mar Hicks (2021) points out how corporations of the technocratic class continually 'tell us to trust them and repeatedly assure us that the tech industry will police themselves and fix their own mistakes. Unfortunately, as a historian, I can tell you this never works' (2021: 21). But if populations and regulations are 'entrusting so much of who we are to large companies, they should entrust us with a sense of where they're keeping it all, and what it looks like' (Blum, 2012: 240). Instead, conditions of trust obfuscate the workings of the data ecosystem.

Market-driven narratives perform and repeat themselves through time, perpetuating the same injustices. For example, Safiya Noble (2018) demonstrates the racialized and gendered issues with trusting in the neutrality of a company such as Google whose business model is advertising yet serves information in the manner of public institutions like libraries. Privatization and profitability have become the guiding factors in determining informational reality and digitalized social structures that define people's everyday lives.

Michelle Murphy (2016) writes of the economization of life, the formation of 'hegemonic liberal economic imaginaries' (2016: 116) that value and shape lives according to risk, speculation and return on investment. While affective and collective, this imaginary, which Murphy labels a phantasmagram, 'does not undo objectivity; it instead adds the trust and belief that animates numbers, or the fear, anxiety, paranoia, or hope that orients and motivates facticity' (2016: 25). Similarly, Geeta Patel (2006: 28) describes the discursive formation (what we could call performative constitution) of 'risky subjects' through the valuing and economization of human lives, the fusing of identities and bodies (like gender and sexuality) with capital, and the naturalization of financial terms of risk and safety as ways of conceptualizing and describing aspects of life.

These financial discourses can be seen in the increasing platformization and ranking systems of already exploitative industries, typically service industries, transport, or small-scale commerce. Individual employees (or non-employees in the case of many platforms who deny even that recognition to those from whom they profit) are constantly valued by customer rankings, relying on 'trust transfer' (Teubner et al, 2019) from a platform (conceived as a combination of the brand and technology) onto the provider. As this trust is not readily transferred between platforms, individual providers must rely on the reputation of specific platforms. This exacerbates the Yelpification of everything, with more complex relationships of trust that position the platform as mediator and manager of trust between other parties (Hawlitschek et al, 2016). The platform economy discourse seeks to monopolize trust not only in certain products, providers or brands, but in the idea of the platform itself.

Even responses to corporate power often fail to escape economic imaginaries. The idea of 'data trusts' acting as 'power brokers' to mediate the use of data by balancing public benefit and individual rights of data subjects (Lawrence, 2016; GPAI, 2022) echoes and amplifies an asset-based approach. While these methods might be critical of top-down ownership/ governance (Delacroix and Lawrence, 2019), if they do not redefine what data is and who it is for, they remain committed (often less than critically) to legal and financial priorities (Raghunath, 2021: 202). These priorities entrench specific conceptions of trust as a matter of individual choice that normalize and legitimize potentially extractive data practices through negotiation of value, rather than challenging the underlying power structures that define the values within which the uses of data are defined. If critical work cannot escape financial terms, then how can it escape the terms of financial contracts and systems of power?

According to economic values and imaginaries, more is always required. More value, more money, more legitimacy, more trust, more metrics, more data. Quantified economization is not simply about volume of data but reach, coverage, a social equivalent to market penetration. Engin Isin and Evelyn Ruppert (2019) write that 'arguably, no kingdom, state, empire, government, transnational or global organization or corporation has ever held such command over the production, storage and analysis of data' (2019: 227). Companies like Facebook or Google have more comprehensively datafied society than government censuses. Communication and media companies like Orange and Twitter supplement government and public data to fill in the gaps in initiatives like the UN Global Pulse. Mobile phone networks and Internet infrastructures replicate the privatization of (for example) energy or transport infrastructures, while the data ecosystem operates according to logics of colonialism, a 'data empire' (Isin and Ruppert, 2019: 207).

These processes enact an 'accumulation by dispossession that colonizes and commodifies everyday life in ways previously impossible' (Thatcher et al, 2016: 990). Tech companies act as, and yet beyond, new imperial powers. They extract trust and legitimacy from formerly colonized governments through the promise of functionality and resilience, and from colonizer governments through appeals to humanitarianism and rights.

Across these contexts, economic narratives drive solutionism as a response to national and global problems. For example, in 2022 the CEO of surveillance company Palantir publicly defended the company's profiteering and normalization of algorithmic surveillance in the context of the Russian invasion of Ukraine with disingenuous appeals to history and rights. As Mbembe (2019) writes, 'the new technologies of destruction are less concerned with inscribing bodies within disciplinary apparatuses as inscribing them, when the time comes, within the order of the maximal economy now represented by the "massacre"' (34). Lives – whether during their span or in their end – have become numbers in the power games of corporations, resources to be extracted to maximize profit.

## Corporate culture

The drive towards constant economic growth in technology industries (and the discourses of technology-supported growth in other sectors) requires continual adoption of new technological objects and systems. This is one area where trust is particularly mobilized. Corporations direct large amounts of money and effort to support trust as adoption. For example, the 2021 Huawei Trust in Tech Summit – themed around 'rebuild, reunite, reset' – featured many talks set oddly and awkwardly amid lavish virtual backgrounds, like stacks of oversized books or strange alien-technology landscapes. This can be seen as attempting to normalize the prevailing visual narratives of virtual reality systems and promote adoption of 'metaverse' discourse, while positioning Huawei itself as an apparently trustworthy leader in this imminent future. These summits and marketing campaigns frame issues as being just out there in the world waiting to be found and exploited, like presenting solutionist responses to removing microplastics from the ocean while ignoring the technological and economic causes of such problems.

Sliding across specific answers to specific problems and the broader narratives of trust and adoption of generic technology, technological solutionism is hailed as the answer to any problem, particularly problems that are themselves human–made and technologically mediated. The discursive sleight of hand that trustification represents can be seen again here in the racket of creating solutions to problems you yourself have contributed to, like Google making a big show of using automation to increase energy efficiency in its data centres. This initiative was used to promote its AI developments, while eliding the continuing contributions of big tech to environmental damage. To resist the use of trust as a smokescreen, we must keep highlighting the entrenching of corporate values in technology discourses. We must keep asking questions about equity, justice and the redistribution of power.

The corporate disregard for the social value of trust and mistrust, reducing it to another metric on the business dashboard, can be seen in the manipulation of discourse and disdain for those from whom trust is extracted. For example, Mark Zuckerberg responded to questions of why people were uploading so much personal information to Facebook by saying 'I don't know. They trust me. Dumb fucks' (quoted in Hicks, 2021: 16). This demonstrates the manipulation of sociotechnical assemblages, sliding between trust in a system, an interface or an individual as a stand-in for expansive networks of data extraction.

Anna Lauren Hoffmann, Nicholas Proferes and Michael Zimmer (2018) demonstrate the performative constitution of Facebook discourse around defining the platform as a utility and social infrastructure, defining of users from individuals, families, communities to the entire world (as well as businesses who remain the real customer of Facebook), and defining what it means to listen (to users' feedback rather than the mass surveillance practices that observe and store everything said on the platform). Marina Levina and Amy Hasinoff (2017) similarly demonstrate how claims of solving global problems are mobilized by platform discourses to push efficiency narratives as justifications for offloading decision making to algorithms embedded with a Silicon Valley ethos of datafication, securitization, individualization and economization.

The culture of corporations is replicated across their internal culture and the way they embed themselves within society. For example, the Data and Trust Alliance is a coalition of companies (including IBM and Meta, Mastercard, Deloitte, Walmart and Starbucks) who proudly measure themselves in terms of revenue and number of employees. They perform the role of a gamers' guild in the gamified world of corporate competition. These games have a longer history in companies through practices such as audits, a quantification of legality that converts superficial appeals to company ethics into an exercise in corporate-washing social responsibility, gamifying issues like justice to force trust and thereby legitimize ongoing exploitative practices.

Employee-based resistance can make headway against these practices. Google employees protested the military-based project Maven with partial success. More recently, Alphabet and Amazon workers have been unionizing for both labour and ethical reasons. But they are confronted with well-financed and increasingly datafied mechanisms for union-busting that use precarity and massive power asymmetries to manage internal risk and enforce trust in the company's own discourse and leadership.

Sophie Waldman, Paul Duke, Rebecca Rivers and Laurence Berland (Scheiber and Conger, 2020), and later Timnit Gebru and Margaret Mitchell (Johnson, 2021), were all fired by Google with little to no notice for standing up against discriminatory systems and practices. They represent groups already marginalized at the company, such as women, queer, trans and Black employees. This treatment of employees asking critical questions demonstrates the lengths to which major corporations will go to constitute their public image. Companies preach maxims like 'don't be evil' even as their actions suggest otherwise. The web of internal and external discourses must be maintained to sustain the illusion of trust and thereby the legitimacy of the business model.

Organizations like the Data and Trust Alliance view the datafication of all businesses as inevitable and desirable, a total transition to a data economy where every business is a data enterprise. Ever-increasing adoption rates are their main KPI. The alliance claims not to influence policy. Instead, they seek to bypass regulation through metrics and adoption, controlling the discourse and entrenching their designs to normalize their aims and uses of technology ahead of regulatory action. Initiatives like removing bias in workplace algorithms emphasize trust in their quotes from member CEOs, backed up by measurements of bias improvement. The product is not a less biased technology, nor even a more effective one. It is a more effective (aka lawsuit-proof) business.

Trust is increasingly seen as a narrative that corporate culture must perform, playing a more prominent role and acting as a priority for corporate leaders (or at least their public image). For example, a report by business research and analysis company Forrester predicted that 'leading companies will seize trust to benefit the planet, empower the organizations and individuals they serve, and seize the opportunities presented by new digital models' (Budge and Bruce, 2021). Again, the appeal to global issues is surrounded by language of extraction and seizure, a data colonialism in the service of economic growth. The same report predicted new corporate roles of Chief Trust Officer, echoing the existing role of Head of Trust and Safety (combining two very different angles) at organizations like Facebook.

These roles, aimed at bridging the gap with civil society to address (or at least counter the visibility of) the harms of platforms, have already been criticized for mixed results.

This is because there is often a fundamental incompatibility between business aims and the level of structural change required to address harms in practice (Ganesh and Moss, 2022). There is an underlying tension between the corporate obsession with growth and accumulation, and more socially embedded values of justice. Addressing the harms of the former cannot be metricized away, for it involves a deep challenge to the prevailing technology discourses that have thus far normalized corporate aims and supported extractive practices.

## Innovation, innovation, innovation

One of the main discourses that supports the corporate exploitation of technology, and of people through data, algorithms and other technologies, is innovation. Bart Nooteboom ties trust to innovation (2013) with a relation of 'cognitive distance' (2002: 25–6). This provides a similar balance to trust–mistrust relations, bringing together legitimacy for action with scrutiny and accountability. Problems arise, however, in the discursive trend towards seeing innovation as an end in itself. Echoing the prioritization of constant economic growth, constant technological innovation has been positioned within industry, government and academia as the support for solving society's problems (or at least making certain people exceedingly rich and powerful in the process). This carries substantial risks for society, such as the social function creep of companies and the entrenching of inequalities.

Function creep of specific systems is often entangled with dark patterns to increase invasive additional permissions for surreptitious and escalating surveillance. It is also seen in a wider social sense. When sociotechnical discourses normalize platforms as social infrastructure, those platforms shift the trust they have extracted in one context (like shopping or online searches) to another, often one with higher stakes (such as digital identity or health).

The biggest giants like Alphabet, Amazon and Meta are particularly active in this regard. They buy out start-ups or rivals within their own sector, like Facebook acquiring Instagram to control and connect the more visual and younger social media platform to their existing userbase. They also expand into additional technology areas, like the shift from Google to Alphabet or Facebook to Meta that represent these broadenings of the scope of their business. They also apply existing platforms' systems, data and logics to new areas, like Amazon or Google taking on health through a combination of personal fitness devices and seeking to embed national health advice or services through their own search and assistant systems.

These companies aggressively expand into areas with heightened social risk. The moves represent a generic view of technology and trust. Innovation is constituted as a flat, neutral term positioning platforms and data as infinitely applicable (or replaceable – fungible, in economic terms). This leads to scenarios where governments and organizations turn to AI developers, platform owners and data analysts to tackle social problems. Those with direct expertise and those directly experiencing the problem are excluded. The story of innovation performs its ability to provide solutions, regardless of the risks to marginalized groups or social values.

The discourse of innovation for its own sake can be seen particularly prominently in issues surrounding blockchains. Blockchain is a technology of discourse, of hype, of the drive towards innovation as a highly monetized and deregulated culture. It is hailed by proponents as the next phase of the Web (3.0). But it is rightly derided by critics for its continued lack of purpose or actual solutions to any practical problem, a form of 'governance by database' (Jansen and Cath, 2021). Blockchains enact a shift from the fungibility of money or the fungibility of infinitely copyable digital resources to the non-fungibility of artificially scarce digital assets. Even fungible cryptocurrency retains a non-fungible element given the way transactions are stored permanently on the blockchain, allowing a tracing of the history of each specific coin and thereby undermining their role as 'digital cash'.

This interplay of fungibility is particularly interesting given the root of the word fungible meaning 'to perform'. Blockchains are entirely performative, constituting value and manufacturing artificial scarcity by speaking it. Money again becomes the proxy for trust in a whirl of speculative investment that merges risk and innovation narratives as a discourse of absent responsibility for the social costs of scams or busts and the environmental costs of the technical systems.

Emerging in prominence from the political project of Bitcoin as distrust in government, surveillance and centralized financial systems, the discourse now not only attempts to generalize trust in blockchains as stores of financial value but also for collective memory – what Angela Woodall and Sharon Ringel (2020) critique as 'archival imaginaries'. Mistrust in one area is transformed into the operationalization of trust as a discourse and site of claims of legitimacy, not only for blockchains themselves but also for the traditional or emerging institutions seeking to take on the roles decentralized blockchains cannot, like upholding rights, guaranteeing investments or mediating disputes (Woodall and Ringel, 2020). This reliance on trust is particularly interesting given the way blockchains are often described as being trustless, purporting to solve the relational social issue of trust through technological design.

But trustless is not the opposite of trust. That would be mistrust, an essential part of the complex, nuanced and contextual process of building trust. Blockchains do not remove trust, they remove context (Werbach, 2018: 29). Trust is simply shifted, enacting 'a change in mediation structures of trust from interpersonal trust mediated by human-based intermediaries to technological intermediaries' (Becker and Bodó, 2020). The trust in people or institutions is replaced with trust in technical objects. But even this elides more complex relationships of trust.

The use of blockchains by people involves various relationships of trust, not only in cryptographic protocols but in user-related boundaries. For example, novices must trust the advice or guidance of those with greater literacy. Web2.0 platformization returns to manage and fill in the gaps of usability in Web3.0 (Marlinspike, 2022). These relations of trust are often based on different, conflicting and implicit interpretations of trust (Jacobs, 2021). In other words, despite the claims of innovation to solve trust, 'some sort of a trusting relationship is a prerequisite for users to adopt blockchain-based systems' (Hawlitschek et al, 2018: 57). Therefore, blockchain discourse can be seen to enact another proxy for increasing power asymmetries or avoiding the need to build a nuanced relationship of trust. Blockchains are a prominent example of anonymization and obscuration of sociotechnical assemblages, propped up by innovation as a justification for trust in the system rather than trust in the people involved.

The discursive role of technology, particularly innovation, is hugely important. Automation of decisions using systems developed by private companies has become worryingly common. The narratives that support such a trend build a culture of innovation, a culture of technologies like AI as the solution to the world's problems. What counts as a problem worth solving is often only those perceived or experienced by the wealthy and powerful. The stakes of this 'algorithmic culture' are 'the privatization of process' (Striphas, 2015: 406), the shifting of legitimacy onto private corporations via their technical systems, built on trust in innovation, and often with little scrutiny or accountability.

The line that 'AI will fix this' has become commonplace among business, government and research leaders. This appears as an appeal to the technical systems in which we are supposed to trust for breakthrough solutions (Bareis and Katzenbach, 2021). But it also carries a discursive turn in which AI acts as a further solution to the problems of existing platforms as well as a response to calls for political action and corporate responsibility (Katzenbach, 2021). The discourses of government, academia and media are already primed to support innovation as an end. The shifting of the narrative back onto metrics of trust in both technical systems and corporate actors supports the ongoing privatization of priorities and values being embedded in sociotechnical systems. The legitimization of these assemblages performed by this shifting of trust between objects and actors is an exercise of the discursive power of trustification. The replacement of trust with financial power has far-reaching influence in setting agendas from regulation to underlying research.

# FIVE

# Research: Setting Terms

If knowledge is power, then the places where knowledge is produced are important sites to examine. They can shape how technologies and discourses develop. Intertwined with the power and discourses of technology companies and states, academia and research agendas also shape, legitimize and entrench trustification narratives. The ivory tower has its foundations in the same need for trust, for legitimacy to support its privileged position, a position not as detached as it is often portrayed. It too relies on stories of evidence, numbers and measurement to extract its power and privilege, from states, corporations and publics. The role of academia in defining technology discourse makes it important to consider how the legitimacy of experts and the results of research are performed as worthy of trust. The global research community (though the term community is stretched here) shifts trust onto methods, institutions and the norms and dominant power structures that shape knowledge production.

There is a long, colonial (as well as racist, heteropatriarchal and ableist) history of reverence afforded to the scientific method and claims of objectivity. Today this has shaped much of how technologies like AI and data are conceptualized and performed. This is significant, as 'methods constitute the things they claim to represent' (Kennedy et al, 2022: 393) while participants, who also shape results, are largely excluded from the same level of consideration (2022: 409–10) even as these methods perform the same values into discourses of society more widely.

In the US, for example, universities played a legitimizing role in sustaining mythologies of settler colonialism, carried forward to contemporary policing and surveillance of any counter discourses that emerge from universities. As Piya Chatterjee and Sunaina Maira write, ' "imperial cartographies" can be traced through the meshed contours of research methods and scholarly theories' (2014). This perpetuates the 'empires of knowledge', which continue to be built on 'racial statecraft, militarized science, and enduring notions of civilizational superiority'. Taken together, this constitutes 'the many different ways white supremacy and coloniality still form the glue for the institutional and intellectual disciplinarity of western critical thought' (Weheliye, 2014: 63). These processes, assumptions and framings embody a sort of quantification in terms of who or what 'counts' as valid for academic freedom or recognition.

As Theodor Porter (1995) writes, 'the credibility of numbers, or indeed of knowledge in any form, is a social and moral one', in which trust is a collective effort and an 'infatuation with measuring [leads] to the neuralisation of concepts' (1995: v). Counting is a homogenizing process. Quantification and datafication as logics of research contribute not only towards a reduction in what can be known, a poverty of thought. They also perform a depoliticization of social issues within research and the context in which research is produced.

## Extracting epistemic legitimacy

The legitimacy of academic power is tied to the trust placed in the knowledge it produces. Whether individual researchers, teams, departments, institutions or the overarching structure of academia, systems of knowledge are often rooted in the same legitimizing discourses. But as feminist thinker Donna Haraway (1988) pointed out, knowledge is inherently partial and situated. It is always created from a particular perspective, in a particular context, and will miss certain things out.

Within dominant academic narratives this is perceived as a limitation. This leads to a denial of the positive aspects of building pluralist perspectives offered by, for example, feminist or anticolonial theories. The partial nature of knowledge, the incompleteness and messiness of understanding the world, creates gaps in trust that dominant narratives feel the need to fill. These gaps, these cracks in the desired image of academia as a monolith of absolute authority, are filled by discourses, and increasingly also by metrics.

The power assertion behind this has a long history of epistemic violence (Spivak, 1988, 24) and epistemic injustice (Fricker, 2007). These in turn underpin the injustices caused by administrative (Spade, 2015) and data violence (Hoffmann, 2018) already discussed, exacerbating colonial heteropatriarchal control over epistemic legitimacy. For example, within the epistemic violence of facial recognition databases we can find 'a "spectral" trace of a past colonial crime' (Delgado, 2021, 71), a crime of exclusion and extraction.

Decolonial theory, like postcolonialism, concerns 'the broader politics of knowledge production and [...] political developments contesting the colonial world order' (Bhambra, 2014: 119). Through this lens, colonial knowledge is seen as defined by the twin discourses of modernity and rationality (Quijano, 2007). This can be traced back to the earliest colonization of the Americas by Spain and Portugal. Even the names Age of Discovery or Age of Exploration hint at knowledge as an excuse for expansion, extraction and violence. Colonialism was built on and created scientific and technological achievements, performatively constituted as a scientific endeavour. It prefigured the Enlightenment and later developments in colonial violence but also epistemic violence of science, anthropology and other fields.

Miranda Fricker's notion of epistemic injustice is useful in examining issues of trust and legitimacy in knowledge production. Fricker (2007) outlines two main forms of discriminatory epistemic injustice. Testimonial injustice is the prejudiced lack of trust in someone's word, and therefore a discriminatory constituting of epistemic trustworthiness or untrustworthiness that become proxies for ethical trust and moral trustworthiness. This is represented by, for example, the denial of lived experience, the historical rejection of qualitative methods by scientific communities, or the lack of trust given to medical patients when describing their symptoms (particularly on gendered, racial or ableist lines). Hermeneutical injustice is formed of barriers to interpreting or describing the world (through, for example and within our discussion here, discourse). This can be seen in a lack of vocabulary for identifying and describing social issues, for example, the denial of victims attempting to define sexual harassment before that term became widely used.

Epistemic power often comes down against positive moves in opening up discourse, performatively aligning those who define existing problems as creating those problems. The need to resist this dismissal is what Sara Ahmed describes in the academy (and society more widely) as 'when I describe a problem I become a problem' (2017). Again, this practice is often steeped in colonial, racist, heteropatriarchal discourses of epistemic control. Others (Hookway, 2010; Coady, 2017) have expanded discriminatory epistemic injustice with distributive forms of epistemic injustice. They raise issues of access to appropriate information that are intertwined with the issues of trust and discourse that Fricker outlined.

The performance of trust in dominant methods and narratives that sustain epistemic injustice works to close off challenges from alternative forms of knowledge. These challenges can be found in the use of decolonial theory to develop more ethical and just AI (Mohamed et al, 2020), or the potential for alternative knowledges like Ubuntu philosophy to reconfigure the terms and conditions of AI (Mhlambi, 2020; Birhane, 2021a). Such wider critical perspectives demonstrate the relational and contextual aspects not only of AI research but of the frameworks of knowledge within which they are embedded, epistemologies that are largely defined by dominant colonial academic narratives.

José Medina (2007) emphasizes the importance of 'shared normative expectations' around speech and discourse as 'a kind of trust' (2007: 41) that is essential to 'genuine sharing of perspectives' and 'meaningful conflict, disagreement, or debate' (2007: 42). Control over who speaks, the act of defining which forms of knowledge or speech are legitimate, is a performative gesture that defines the landscape and limits of knowledge. This can be used to promote wider interpretations and engage with alternative frameworks of knowledge, or it can be used to collapse contexts and cultures into dominant epistemologies that perpetuate injustices.

The contemporary manifestation of the colonial roots of knowledge production can be seen more widely in the tendency towards data colonialism (Couldry and Mejias, 2019) as appropriation and dispossession (Thatcher et al, 2016), which leads to 'the violent imposition of ways of being, thinking, and feeling' (Ricaurte, 2019: 351). This sits within the colonial operation of computing more generally that decontextualizes knowledge through, for example, an overwriting of cultural epistemologies with logics of abstraction and universality (Irani et al, 2010). Against colonial framings and practices of knowledge and technology production, we can look to what Quijano (2007: 117) calls 'epistemological decolonisation', to which Couldry and Mejias (2019) add a call to reject the supposed rationality of colonial appropriation in constant data collection.

Decolonization acts as a challenge to the underlying structures of knowledge that Paula Ricaurte describes as 'data epistemologies'. These are 'based on three assumptions: (1) data reflects reality, (2) data analysis generates the most valuable and accurate knowledge, and (3) the results of data processing can be used to make better decisions about the world' (2019: 350). For Ricaurte, knowing, being and sensing are integrated within colonialities of power. Discussion of these forces tends to focus on the power of corporations, but we can also usefully extend this conversation with the contextualization of academic knowledge production as performing not only colonial logics.

This 'decolonial turn' is itself further enhanced in connection with critiques of wider logics of injustice such as race (Couldry and Mejias, 2021), class, gender, sexuality and disability. It is important to stress these multiple and intersecting forms of injustice embedded within data universalism, 'the original sin of Western interpretations' of knowledge (Milan and Treré, 2019: 325). Dominant frameworks of knowledge produced by centuries of colonial academia persist today. They require critique across multiple dimensions of oppression that have underpinned the way technology discourses have been framed and embedded within systems of power.

A particular site of academic power structures and influence, especially in technology-related fields, is the lab. In their history of the laboratory, Maya Livio and Lori Emerson (2019) propose a feminist critique for rethinking what a lab might be. They begin from the anatomical theatres, apothecaries and informal kitchens where experimentation took place, spanning educational, business and personal settings, before discussing tensions of gender and class in the transformation of these spaces into labs. This is followed by the industrialization of laboratories in the nineteenth century, merging with the colonial roots of scientific methods.

The colony often served as a laboratory (Tuhiwai Smith, 2012), replicated in the racialized treatment of populations such as the displacement of indigenous, Black and poor populations to build large-scale labs as well as the direct experimentation on those same populations (Yusoff, 2019), a further necropolitical underpinning of science and technology as a method. Livio and Emerson (2019) pay particular attention to the MIT Media Lab that has become hugely funded and influential within technology disciplines. Their response is to consider a series of feminist principles: complexities of access; epistemic assumptions; labour; hierarchies; safety and affirmation; and environmental concerns. Epistemic issues are particularly important within a feminist critique for understanding the constitution of expertise as well as the assumptions of objectivity and the entrenching of certain methods and knowledges.

## Objectifying knowledge

Addressing objectivity is an important part of tackling issues of legitimacy and trust in academic (and corporate or state) research. Naomi Scheman (2011) highlights how belief in the objectivity of scientific research relies on internal norms of the research community that are reliant on practices of trust-grounding, problematizing the issue of 'epistemic dependency, and the implication that knowledge rests on trust as much as it rests on such epistemological staples as perception, reason, and memory' (2011: 214). But, as Scheman suggests, 'what we label "objective" has actually to be worthy of our trust and the trust of a diverse range of others' (2011: 210). This highlights the importance of collective and relational trust, not just of a specific individual in a specific context ('who do I trust') but of building trust across different communities and populations as a support for the legitimacy of research claims as something that can be generalized.

This leads us to a feminist critique of objectivity itself, the problematic construction of scientific methods (and the results they produce) as being detached and impartial. Donna Haraway wrote that 'the knowing self is partial in all its guises' (1988: 583). This critique confronts objectivity and completion narratives, the constitution of problems as universally solvable apart from the contexts in which they operate. Instead, feminist thinking insists that we situate knowledge in the context in which it is produced and applied, focusing on communities rather than individuals – whether that is the narrative of the lone researcher as 'leading expert' or whether it is the individual supposed to believe in this objectivity. Complex social problems are often fundamentally unsolvable, but the pressures of funding and publishing mean that academia needs us to think they can be, echoed in innovation and technological solutionism narratives and sustained by trust in the legitimacy of scientific discourses.

Against these discourses, feminist objectivity is not universal but a combination of partial perspectives (or standpoints or positionalities). As Catherine D'Ignazio and Lauren Klein show within discourses of absolute trust in data within research, 'the quest for a universal objectivity – which is, of course, an unattainable goal. The belief that universal objectivity should be our goal is harmful' (2020a). And yet universal objectivity supports discourses of generic innovation and technologies that can be applied to any social issue. The colonial and patriarchal discourses of academia continue to push scientific objectivity as a root of trust that entrenches hierarchies and provides a discourse of legitimacy.

Technology discourses supported by the epistemic injustices of (academic) knowledge production have been turned around onto the production of knowledge in academia itself. Today, exploitative mechanisms of metricization and trustification are commonplace within research and teaching, a computationalization of academia. But, as with many of the trends towards datafication and trustification, this is not entirely new.

Mechanisms like peer review (as well as wider practices like the disciplining effects of discipline-specific copy editing) are linked to traditions of 'academic legitimation and accreditation' (Hall, 2016: 21). These systems further entrench certain historical biases linked to the expression of knowledge by codifying specific, often by field or geography, ways not only of producing but of presenting new knowledge in order to legitimize it. This is an entire system designed to perform trust in its own processes and products, to perform its own authority.

Similarly, Library and Information Science has seen a tension between the wide range of practices that acknowledge the need for different and inclusive approaches to knowledge, and the push towards standardization in increasingly monopolistic epistemic framings and platforms. Databases like WorldCat have for decades attempted to collect all knowledge under one bibliographic roof, a digital manifestation of Borges's Library of Babel. Concepts such as 'authority control', which seeks to define entries under common labels, have spread to mechanisms like ORCID, which apply a similar logic to academic authors. The standardization of names attempts to attribute due recognition. But it also carries with it a globalizing homogeneity that sidesteps issues of different writing systems or different naming conventions in other cultural contexts, not to mention the potential harms of linking trans authors to dead names (a name given to a trans or non-binary person at birth that no longer fits with their gender identity). Cultural and contextual specificity is erased by exacerbating the dominance of Anglophone conventions. This also shifts from identity to identification in the conversion of authors into numbers, all in the name of consistency.

Increasingly, metrics and datafication play a role in defining the process of doing, distributing and accessing research. Computational logics and discourses have come to dominate these definitions. This has been in part due to the hacking together of informational infrastructure by Computer Science departments with the technical resources and skills to set up and normalize metrics-based systems and logics.

For example, the DBLP database of Computer Science research has been developing since 1993, over a decade before the platformized approaches of Google Scholar or Mendeley, and now holds information on millions of papers and articles. The quantification extends beyond measures and comparisons such as citations or h–index, towards questions over what 'counts'. Google Scholar, for example, tends to be overly inclusive in what is added to an author's profile. Discipline-specific databases like DBLP have been built up from records of specific authoritative venues, with the counter problem of excluding multidisciplinary work or authors who straddle traditional disciplines. In either case, the process is increasingly automated. DBLP creates pages for authors automatically when they publish in certain journals without including publications in venues that fall outside the designated field of 'computer science', while Google Scholar maintains the aggressive scraping familiar to its parent search engine.

Metrics in academia are linked to automation. Evelyn Ruppert (in Neilson and Rossiter, 2019: 193) writes of the 'crafting of automation', the role of metadata in establishing trust in data. This extractive process is increasingly automated according to standardized rules of quality and governance, performing its own validity leading to more automation. In academic research, this can be seen in the metrics that establish the authority of certain works or authors and trust in the journals and platforms that host or grant access to such knowledge. The metadata of research has taken on a powerful role that raises the visibility of some research (and researchers) over others.

These logics echo the performative constituting of knowledge to design technologies that frame norms of future knowledge production. For example, from MATLAB to Google Search, key tools for automating the ways we define, access and generate knowledge stem from doctoral theses exploited by industrialization. Postgraduate students and precarious research (and teaching) staff on temporary and/or fractional contracts are an exploited workforce individually consentified into the system. To reach the dangling promises of stable employment – in the form of mechanisms such as tenure and tracked through publications, citations and grant income – entails a tremendous amount of hoop-jumping to meet the ever-escalating requirements of institutional metrics.

An example of metricized academia at scale is the UK's Research Excellence Framework (REF), a seven-yearly exercise that measures and ranks the quality of research at every research institution in the country. Much of this is based on a supplementary rating of peer-reviewed research outputs by subject area. The results define sizeable government block funding for the following seven years, not to mention ranking reputation between institutions.

Derek Sayer, an outspoken critic not of the metrics themselves but of the decisional processes by institutions in the REF, discusses the designs and practices that encourage gaming by institutions, with academic staff as the resources to be exploited. He writes that 'one thing a regime like the REF does very effectively is to create a perpetual climate of anxiety, especially when many of the rules of the game are not known' (2015: 78). He also highlights the power sitting behind rigging the rules of these games by already powerful institutions and figures who can influence the design of the next iteration of the exercise (2015: 102). This normalizes abusive hiring practices and places burdens of metrics-chasing onto individuals in service of the institutional game.

Sayer also notes the slippage of 'peer review' to 'expert review' (2015: 29), a conversion of academics reviewing each other (a process not without problems of its own) into a review by a select few deemed expert enough by policy makers. Peers are in principle equal, whereas the term experts implies and creates hierarchy. Slippage occurs across the assemblages of knowledge production. We see slippage from individual outputs or individual academics to research environments (especially department or faculty level) and institutions. The goals of departments and institutions are pressured downwards, placing the burdens on individuals to deliver, and justifying the extraction of knowledge from research staff.

The REF also entrenches historical biases. It remains the Russell Group of (old, elite, research-focused) Universities that dominate the scores and therefore the funding, being able to play the games of both quantity and quality where resources beget more resources, all legitimized by the metrics discourse. These 'research-intensive' universities are also increasingly becoming 'audit intensive' (Holmwood, 2016). Internally and externally, they emphasize metrics of publication rates, impact factors and grant receipts, performing the cycle of metricization that keeps institutions like Oxford University on top.

There are also higher-level considerations of funding decisions made based on metrics, separated from peer review and expert review. This legitimizes the ongoing dominance of historical powers, often those elite universities with colonial histories. All the while, these games impact on the careers and research goals of individuals. Academics must internalize the needs of the metrics game, and trust that their efforts will be rewarded through recognition and promotion despite the gaming of employment practices by universities to place the institution in the best position for a higher ranking at the expense of other considerations. Metrics enable already powerful institutions and systems of knowledge production to further extract trust in the validity of the knowledges they produce, solidifying their legitimacy to claim expertise and wield power over knowledge.

## Performing expertise

The extraction of legitimacy from academic labour via metrics helps constitute the role of academic knowledge production within society. It is therefore important to consider the wider discourses that trustify publics in the legitimization of academia as an authoritative source of knowledge in society. Boswell (2018: 42–3) positions authoritative truth as a basis for political trust, and this highlights the need for academia's role to be supported by trust that the knowledge it manufactures is indeed truthful and authoritative. This authority is always situated within broader geopolitical contexts, such as the economization of populations and transnational 'experts', expanding beyond academia into government, NGOs and consultants within Cold War and postcolonial settings (Murphy, 2016). The shifting onto 'experts' performs a specific role and power relation within society, one where academia still seeks to maintain a monopoly.

The discursive practices employed to support this go beyond a sense of authority into measures of accuracy and perceptions of sincerity as roots of trust (Fricker, 2007: 76, 111). By performing not only expertise but also sincerity as proxies for moral and ethical trustworthiness, individual 'experts' performatively constitute their legitimacy to speak. The discourses of academia and technology development intertwine in the narratives that experts can (and should be allowed to) improve society, extracting a political form of trust in intentions and abilities.

The performance of historical discrimination over who counts as such an expert reiterates and exacerbates epistemic injustices, particularly those along racial, geographical and gendered dimensions. Such injustices add onto wider issues like university pay gaps, often rooted in the quantification of mechanisms like recruitment, promotion and tenure that favour traditional models of credibility and hierarchy. As these narratives are performed over and over, measurement becomes the goal. This reproduces historical inequities in who has resources to conduct research and who then gets to have their voice heard as an expert, who gets to shape knowledge and the discourses that surround it.

During the COVID-19 pandemic, institutions like Oxford University and Johns Hopkins University positioned themselves as authoritative voices on research, covering topics like vaccines and tracing apps or statistics respectively. These two are also the longest running research institutions in their countries (UK and US), and between them represent both old and new colonial powers of knowledge and technology. Oxford was also embroiled in debates over contact tracing apps and competing academic discourses around trust in technical systems, specifically the interactions between cryptographic protocols, real-world implementation and wider sociotechnical power structures (such as Google and Apple's decentralized model versus the UK government's planned centralized system). Competing consortia of institutions authored public reports or letters to extoll the virtues of their preferred system and discredit the opposing view, seeking to elicit influence with policy makers by extracting a nebulous sense of legitimacy from public academic discourse. I discuss this in more detail in the first case study (Chapter Seven).

José Medina (2012) outlines how 'excessive attribution of credibility' contributes to epistemic injustice in broader social contexts (especially racial and gendered social inequalities) (2012: 59). By shifting credibility onto one place (institution, person), this exacerbates the difference when credibility is denied or inappropriately attributed to marginalized voices. This is evident in the dominance of elite, colonial universities in pandemic research but also in discourses across all fields and particularly technologies.

Credibility is performed through a combination of metrics (funding, publications) and access to narratives (being the go-to expert for mainstream media). Those wielding excessive credibility perform a judgement of being 'more worthy of epistemic trust' (Medina, 2012: 63). The combination of deficits and excesses of credibility exacerbates existing dimensions of marginalization, contributing to 'an epistemic injustice that is grounded in a comparative social injustice' (2012: 66). The iterative performance of these comparative power dynamics increases vulnerability in self-trust for marginalized groups. Trust is shifted away from individuals and onto institutions of power. This further restricts access to credibility, discursive power and the shaping of knowledge in the world.

Knowledge is social, and the use of credibility metrics to extract legitimacy for specific individuals, institutions and agendas is firmly rooted in existing social injustices. Medina (2007) had previously shown the way that 'marginal discourses that speak on the borders or at the limits constitute a risky speech that has a dangerous and precarious life' (2007: xiv). He emphasized the need from within and without academia to 'break the silences' produced by epistemic injustice, but that this entails 'the arduous process of contesting the normative structure and power dynamics of discursive practice' (2007: 193). Credibility acts like a resource in the games of academic recognition and hierarchy. The stakes of these games are control over the discursive framing of knowledge. We must challenge the way proxies such as credibility support the underlying logics of metricized trust that trustify knowledge. The currency of credibility extends to inequalities of influence over shaping future research.

## Constituting research agendas

The various metrics that establish the legitimacy of dominant systems of knowledge production perform discourses of power that shape the underlying logics and dynamics of technology and related fields. This finds its expression in the relation between academia and other sectors. For example, research from Oxford University that sought to 'measure bias' in algorithmic decision-making systems claimed that we 'need metrics that reflect our value systems' (Oxford University, 2022). The university proudly shared how Amazon adopted the modelling tool in the SageMaker Clarify system.

This needs problematizing. Firstly, there is a significant gap between metrics that reflect values and embedding those values themselves in the full gamut of the design and implementation process. Secondly, the SageMaker Clarify system functions during data preparation, after a model is trained, and during deployment. This conveniently avoids dealing with the deeper injustices of inequities in design or the more fundamental question of whether a particular system should be produced in the first place. Thirdly, the mutual relationship of credibility entrenches social power as Amazon uses the research narrative to trustify its algorithms by extracting legitimacy from the proxies of the institution's credibility and trust in the epistemic framing.

Further questions remain over whose values are embedded within the system, and who decides those values. Is it UK law (where Oxford is based), US law (where Amazon is based), the laws of anywhere it is used, the more nebulous social values of those countries' cultures, the values of Oxford University (or the research team), or the values of Amazon as a private company? Beneath the mediatized excitement of collaboration to performatively tackle social problems, these problems are constituted and defined in ways that support the underlying academic-corporate values. This framing not only slides injustice into the more packageable problem of bias but labels this problem fundamentally solvable.

Similar logics can be seen in, for example, the annual Stanford Human-Centred AI Index report, which metricizes AI research outputs by volume of papers in conjunction with economic aspects like national and private funding by country. This performs colonial and capitalist narratives as defining measures of research value. It also emphasizes the literal quantification of research echoed in the obsession with volume in discourses surrounding 'big' data and 'large' language models where authority is proxied through quantity.

Global research agendas are maintained by these metricized approaches and the colonial-capital logics that underpin technology and wider research. The existing disparities in research funding and influence were highlighted during the COVID-19 pandemic as funding bodies rushed to support urgently needed research while institutions competed to take a bigger portion of these resources. This resulted in the extractive and exploitative trustification of the Global South by the Global North, through disparities in research output relating to vaccines combined with colonial logics of ownership and production. For example, vaccines based on patents held by private entities in the Global North were manufactured in countries like India and then hoarded by wealthier governments such as the UK.

A live global tracker of pandemic-related funding (UKCDR, 2022) showed the dominance not only of medical research (as might be expected) but also of technological responses.

Of social aspects, the largest was WHO sub area 9a (Public health focusing on acceptance and uptake, which can be read as trustification proxied by adoption), with over 1,000 projects totalling over $300 million. Next was WHO sub area 9c (Media, including issues of trust), with over 500 projects totalling almost $150 million. Next in monetary value was WHO sub area 9d (Engagement – relevant, acceptable and feasible participation practices), with over 150 projects totalling over $100 million. Ethics was by far the lowest funded, with less than $50 million allocated (all figures based on the time of writing). Almost a quarter of all projects were US-based, and the US and UK dominated the funding (almost half were in these two countries, leaping to over half if we include Canada). This Anglophone dominance perpetuates linguistic and colonial narratives.

In part this could be down to where funding is coming from. These colonial powers have well-established research funding mechanisms that were able to rapidly divert significant resources to pandemic-focused projects. However, in the context of the research imperialism that increased during the pandemic, this does not absolve the need to situate research within affected communities and incorporate global perspectives. Social science (WHO area 9) projects were more evenly distributed, though the UK and US still dominated, followed by Canada and the rest of Europe. Ethics (WHO area 8) projects were sparser in general, but the US, UK and Germany still dominated. Only three projects were based on the African continent (Nigeria, Kenya and Zambia) with none in South America.

This demonstrates the specific narratives that define ethics research in medicine and technology. This form of ethics supports rather than challenges dominant narratives. It acts as another source of measurable legitimacy by promoting ethical boxes to tick rather than deeper sociopolitical engagement with issues of global injustice. The sliding of trustworthiness onto supposedly objective ethical qualities of technical objects performs the role of obfuscating the ways those technologies are situated within social and political framings of knowledge and the wider discourses that embed social values and injustices.

The assemblages of knowledge production, and the ways they are situated and performed within wider power relations of sociotechnical assemblages, perpetuate dominant inequitable discourses across such assemblages. Epistemic infrastructures sustain historical narratives spread by elite institutions playing games of influence and power. Meanwhile, individual researchers are expected to compete in their own games of recognition to situate themselves within the hierarchies such infrastructures produce. Trust is extracted upwards from each level through the metrics that govern such games, legitimizing the power structures that perpetuate epistemic injustices of technology discourses. These discourses are legitimized within society through the extraction of trust in the form of the attribution of credibility for research expected from populations at large. Academia performs a specific role of influence and power built on the ability of quantified knowledge and metricized knowledge production to convey the legitimacy of expertise to wider publics.

# SIX

# Media: Telling Stories

The stories we tell about technology are important. The legitimacy of the process of trustification is built on stories of quantification, extraction and power. These stories are spread by states, companies and academia, but they find their expression repeated, performed and measured in the media.

Narratives and discourses mediate and are mediated by technology companies and platforms. They are performed on platforms about platforms. From persistent forms such as the press, advertising and film to each new social media company, media act as the stage and frame for constituting our collective discourse of technology. Measured in clicks, likes and shares – or on a different level in reach, ad revenue and market share – trust is extracted in and through media. Narratives are built and challenged around the conflicting and shifting loci of trust in media assemblages. The apparent depoliticization of quantified trust belies the highly political nature of the way such systems render trust little more than a metaphor for operationalizing the engagement of populations to legitimize power. Every minute spent scrolling performs the conditions of trust in media regardless of whether trust even enters the equation.

Examples of these shifting narratives are everywhere, each one contributing to the constituting of wider discourses of technological solutionism in different ways, in different contexts, to different audiences. Chris Gilliard and David Golumbia (2021) write of the differences in power and privilege that can transform the perception of surveillance technologies into both a tool of oppression used against the marginalized or a luxury item for the wealthy. Devices that collect obscene amounts of personal data are repackaged as high-end consumer items desirable for features that would rightly scare others, all based on the perception of branding and media that justifies the costs (both financial and social).

Even when companies like Apple make grand gestures about user privacy, it comes at a cost few can afford and still places data into the hands of that one trusted luxury brand. The same technologies are felt differently by different groups. Embedded within sociotechnical systems and discourses, these affective qualities create wildly unequal effects. A 2009 Apple iPhone advert popularized the phrase that whatever you need, 'there's an app for that'. The way this became trivialized as a joke conceals the combination of lifestyle, brand and utility narratives that normalize a data economy as the dominant understanding of complex informational ecosystems.

This story is driven by the generic development of specific functionality. Menstrual cycles, exercise, productivity at work, a child's progress at school: all are invasively tracked. Lives are quantified and data extracted as part of a generic data discourse through apps created by those without expertise in the specific areas they claim to solve. Standardization is generalization is homogenization.

Trust in these different settings is extracted through standard measures like downloads and user ratings, supported by endorsements, paid-for advertising (often on other apps) and the freemium mentality. This combines the 'personal data in exchange for services' logic normalized by the likes of Google and Facebook with the continual push to extract money directly from users through mechanisms like artificial scarcity and financial barriers. Techniques that promote constant engagement – the infinite scroll, swipe down to refresh, the gamification of reward – embody the processes of trustification. Engagement metrics performatively construct proxies for the conditions of trust while trying to escape the need to truly build trust in corporate-owned media. Trust becomes implicitly extracted for as long as people keep doomscrolling.

Similarly, from big data narratives where quantity rules, to the mythic status of AI and those who build it, trust is elicited through the apparent objectivity of numbers and the systems that produce and process them. The wider story of technology as the solution for social problems (even those created by the same or similar technologies and the companies who produce them) can be traced back in criticism and historical sociotechnical theories and resistances. These problematic narratives have been given various names: technopoly as a replacement for culture (Postman, 1993); the sublime or myth of the digital (Mosco, 2004); the technological fix for social problems (Volti, 2014); solutionism (Morozov, 2014). Christian Katzenbach (2021) outlines how the technological turn to AI, including its normalization and widespread deployment across issues and sectors, 'is at least similarly dependent on a parallel discursive and political turn to AI' (2), particularly in managing the digital media content that goes on to shape and constitute future discourse.

Technology narratives function by sliding trust from one entity (concept, artefact, infrastructure, organization) to the next. Contemporary media are highly complicit in these conflating and confabulating performances. Simone Natale (2021) goes further to define AI as a technology of deception. Its technical success is based on deceiving humans into thinking machines can act in their place. Science fictional representations and fears are entwined with justifications for funding and development through constant misdirection and redirection of attention and efforts onto new technologies. It is media(tized) narratives that enable a specific technology (or technology discourses more generally) to stand in for specific trust relations in complex sociotechnical assemblages. Across these mediating processes, the discourse continues to be peddled that legitimizes trust in measures and metaphors rather than a concern for people and power.

## Representation escalation

Technology discourses operate not only through sliding trust, attention and meaning across objects and actors of complex sociotechnical assemblages. They also enact a sliding of terms themselves, which become signs for little more than hype and power. Is a 'black box' a technical object that makes decisions inaccessible to those operating it, or the way that object fits into practices that keep its operation secret from outsiders? Is 'cyber' a space, a type of security, a way of thinking, or a culture? Is 'crypto' a field of study, a set of mathematical protocols for encryption, an alternative financial system, a politico–economic movement, or a pyramid scheme? And what is 'AI', really? Terms such as these often function as proxies for tech and expertise always desirable, yet always out of reach.

The diverse range of possible meanings and intentions in what constitutes or counts as 'AI' shapes what questions we ask (and are allowed to ask) about it. How we frame technologies impacts how we frame discourse, and in turn how we can frame critique, how and where we can direct mistrust. For example, Jenna Burrell (2016) outlines three different forms of opacity, which equate to issues of governance, expertise and epistemology. These mechanisms block or misdirect understanding and criticism across the sociotechnical assemblage of what a particular AI includes. Overcoming these barriers requires responses that may be either technical or social, promoting greater media and public understanding while also holding to account the power structures and hype machines that perpetuate harmful or misleading narratives.

Media representations rapidly escalate the value and image of technologies such as AI. The ever-shifting locus and meaning enacts a mystification, an enchantment. From the various forms of opacity, AI and other technologies generate an image of magic, following which it becomes 'unsurprising that AI and more generally electronic computers stimulated from their inception a plethora of fantasies and myths' (Natale, 2021: 35).

These myths include the 'magic of big data and AI' (Elish and boyd, 2018). Religious imagery such as creation memes (Singler, 2020) are used to generate trust and authority, and technologies wear 'the garb of divinity' within what becomes 'a computational theocracy' (Bogost, 2015) that transposes the functional legitimacy of the technical object onto a system of power.

These different discourses of mystification contribute to the continual shifting of the loci of trust, an avoidance of questions and answers. A trail of legitimacy slides from technical objects to designers to regulators to marketers to discourses. This lands back on technical objects in a way that enforces trust in the generalized concept of technology rather than any particular object, system, context or relations. This discourse perpetuates the legitimacy without ground through narratives of mysticism and the purposeful obfuscation of sociotechnical issues.

Across all these questions, these representations, these displays of symbolic power over discourse, trust is located more in these performed meanings than in any concrete functionality. In short, AI and other technologies ask us to place trust in hype: 'AI currently enjoys a profound as well as multifaceted hype that might be rooted in the sort of ambiguity that comes with an uncertain and contingent future' (Roberge et al, 2020: 4). This hype reflects the corporate and economic thinking I discussed in earlier chapters, where risk management and metricized speculation rule decisions with highly social implications.

The spread of such discourses through the media brings in a particular way of thinking that is strongly rooted in metrics of trust such as readership, engagement and the bottom line of advertising revenue, as well as links with corporate and related discourses of power. The normalization of hype as the reality of media representations creates a 'need to collectively recognise the power, mechanics, and sheer volume of it' (Milne, 2021: 119). We need to unpick the ways such hype functions and the effects it has on those subjected to the hype itself and/ or the effects of technologies sold to power using such hype.

Echoing the dominance of trust in AI narratives of government reports and academic papers, news media pushes the same assertion that technology 'ought' to be trustworthy, and to be trusted. Examples abound of articles proclaiming the need for trust in AI or bemoaning a lack of trust as if it is a resource to be provided by the public or advising companies on how to enhance the trust of their brand (either for selling AI or using AI to sell something else). For example, the *Financial Times* published an article on trust in AI as 'essential' (FT, 2021). This is notable due to its authorship by the editorial board, and the fact that access was largely paywalled. There is a cost to gain access to this panel of experts even though they are tightly bound with this (business-focused) publication.

In a similar vein, *Forbes* published two articles in one week on trust in AI (Ammanath, 2022; Banerjee, 2022) as part of their invitation-only, fee-based Business Council posts. Entry requirements for this council reserve access to this boost within the highly metrics-driven world of online content visibility for those who own or run businesses with a certain revenue. Systems (or rackets) such as these demonstrate the gamified aspects of trustification within online media. Trustification imperatives override editorial and journalistic principles or discourses with a game played by the elites using audiences, their clicks, views and reading time (aka potential advertising time), as resources. It is simultaneously a private club that requires trust in its branded exclusivity, and a competition for these elites to win over the masses to trust their brand, their image, their hyped-up technology product over others.

However, though the tales might often be tall, 'it is not all empty hype: the statistics trotted out in press materials and in company transparency reporting illustrate the significant role that automation is already playing in enforcing content policy' (Gorwa et al, 2020: 2). Even though such statistics often rely on flawed measures or false comparisons (Gillespie, 2020: 3), they may well contain within them – or performatively constitute when published in supposedly trustworthy news outlets – a grain of truth about the discursive power of technology if not its functionality. The question of whether an AI system can detect cancerous tumours or generate accurate answers to essay questions becomes subsumed beneath the question of whether it is convincing 'enough', whether cherry-picked results or cases can persuade media, publics and decision makers that state-of-the-art systems can be trusted to function as intended when thrust into complex real-world use. And even when a technology like AI or machine learning becomes 'hardly obscure or arcane knowledge today. These techniques are heavily documented' (Mackenzie, 2015: 431), the performance of a gap in understanding constitutes a truth of sociotechnical opacity where the power if not the technology is obscured or inaccessible.

This displays the 'dual nature of hype', two opposing roles of constituting understanding through performing metaphors, where 'the very lubricant that eases ideas through complex systems is what allows snake oil to make it to market' (Milne, 2021: 121). By embedding both the myth and reality within culture so broadly, these discursive practices further normalize, further extract legitimacy for, further impose the expectation of trust in, a 'power of generalization that currently seems intent on overcoming all obstacles' (Mackenzie, 2015: 433). Media representations, even when relying on hyperbole, metaphors and myths, create a blurred enough truth to enable the sleight-of-hand tricks that speak themselves into belief and thereby perform their own trustworthiness.

The performance of multiple meanings returns here. The inability to pin down a common or accessible understanding of what a given technology is, does or could do constitutes misunderstandings by a wider audience, which often includes groups like policy makers. This in turn entrenches the expertise gap between technological elites and the people whose lives they believe they can 'fix'. This gap creates the conditions in which the mythical and magical qualities of technology become believable.

The whole process performs an enchantment. Words have power, and in the media this generates symbolic power in a more literal sense. Controlling understanding and discourse controls how technology fits into cultural tropes and sociotechnical assemblages. Trust becomes the resource through which this 'spell' operates in material and social ways. We return to issues of power and the control over discourse to extract trust through proxies of mythical representations that justifies the legitimacy of technologies used to entrench systems of power and systemic inequalities.

## Systems and content

Technical systems are also systems of power and systems of social relations. Yet these relations are datafied, gamified and consentified to extract trust and enforce compliance from those within and without the organizations designing, developing and deploying these systems. For example, Ashwin (2014) outlines how trust is an integral part of the production of communities within social and corporate discourses at the level of governing the technical aspects of the Internet. This is wrapped in a narrative of acting 'for the good of the Internet' that embeds political centralization and economic interests (2014: 226).

In this sense, infrastructures are not objects or even technical systems. They are the production of social relations, power relations, and the performing of communities as mutual trust between individuals, the collective, the technical objects and the discourses that constitute a shared vision and direction. But far from decentralization, at the level of discourse we see the high-level game between powerful interests with the technical community as the resource to be trustified. Competing agendas of established centralized power subsume the open collective and any social mission.

Even within communities that define and develop technologies, trust is displaced throughout sociotechnical assemblages. This is only exacerbated when it encounters wider political and social communities and wider media. For example, the Clearview AI facial recognition system deemed illegal in many countries due to its mass extractive non-consensual data collection practices was used by Ukraine to identify killed enemy soldiers during the Russian invasion. These were used not only for internal records but shared via social media. The narrative was pushed that it was to counter Russian state propaganda and find alternative ways to notify the families of the deceased. But it was also entangled with Ukraine's own propaganda efforts as well as the normalization of technological discourses during a moment of crisis.

There is another circularity here. The images are scraped from social media and end up on social media, but there is nothing social about it. Their journey is via a corporate actor with highly questionable practices, as well as the state of Ukraine, passing through different forms of power, passing through different contexts, but circulating around and about people without granting them any agency. The return of the faces to social media in death demonstrates the necropolitical power of systems that datafy individuals. They are reduced to resources in the propaganda games between states that attempt to bypass one another and pass directly back to populations to undermine trust in their own government, as well as the corporate game of shilling illegal technologies to those existing powers.

This collision of state and corporate logics at the expense of the social interests of communities and populations can be seen in copyright. Copyright was one of the first use cases for automated content matching, moderation or removal, embedding economic imperatives and corporate interests (especially those of media production and distribution companies) in the design of systems (Gorwa et al, 2020: 7). These debates continually re-emerge in issues of automating content moderation. The construction of laws by states that prioritize corporate interests has had further impacts on digital platforms by constituting a discursive norm that entrenches the corporate-focused logics of copyright even when states then try to regulate media distributors, particularly with the loophole-seeking practices of social media platforms.

Despite their integral role in contemporary social life, media platforms receive very low levels of trust from their users. In an Axios (2022) poll of US companies' reputations, social media platforms Facebook, Twitter and TikTok ranked in the lowest 10 per cent, with trust among their lowest constituent scores. Other media companies such as Fox also scored very poorly and, while Disney were only slightly below average, this represented a significant drop compared to previous polls. Meanwhile hardware and informational infrastructure companies such as Samsung, Microsoft, IBM, as well as Google, scored very highly, with significant increases in reputation, although trust (along with citizenship) remained a weaker point. Even in quantifying trust, social media platforms are losing their reputational legitimacy, which only heightens their need to extract trust from their user communities and reposition themselves as essential and inevitable social infrastructure.

The design of social media systems is rigged against users' interests. Advertising revenue and engagement metrics reign supreme, and the optimization of these goals creates boundaries on public discourse on such platforms. This can be through dark patterns of interface design or of algorithmic governance, producing conditions that limit user behaviours and discourses. Communities on platforms must 'invest energy into performing "properly" for the algorithm. Invisibility (or its threat) is key to the structure' (Maris and Baym, 2022: 332); meanwhile narratives around these algorithmic structures can lead to mistrust not only in systems but also redirected onto other users and communities.

This performatively sets the stage for what performances are possible. Platforms establish the ways people can speak and act online, they set the context and content of discourse. This has an impact on how communities can produce themselves in such spaces. Quantification emerges as the precursor for scale (of systems, of power, of harms), and Tarleton Gillespie (2020) points out the way 'AI *seems* like the perfect response to the growing challenges of content moderation on social media platforms […] often justified as a necessary response to the scale' (2020: 1, emphasis added) but 'it is size, not scale, that makes automation seem necessary' (2020: 4). Gillespie argues that it is capitalist narratives that must be challenged to reach more effective governance of online social spaces.

The AI solutionist narrative enacts a 'discursive justification for putting certain specific articulations into place' (Gillespie, 2020: 2). This further entrenches economic and data imperatives like automation escalation, constant mass data collection and infinite growth. The increasing overlap between content distribution/moderation and AI performatively constitutes its own legitimacy. Size requires automation, which enables size and the impression of scale. But while the systems might on a technical level be only increasing in size, it is the harms and social effects that are scaling outwards into ever more worrying realms.

Emma Llansó (2020) cites the example of proactive content removal, the prior constraint issue of censorship in advance, a human rights issue of potentially blocking certain people or groups from speaking. This practice merges content creation, distribution and moderation. According to Llansó, proactive content removal has three mechanisms: encompassing more speech; removing procedural barriers; reducing scrutiny. The apparent (self-constituting) need for scale shifts discourse from a human rights-based presumption against prior constraint towards a platform automation assumption that it is a useful tool or scalable solution. Similarly, Gorwa et al (2020) identify problems of increasing existing opacity (thereby reducing scrutiny), complicating social concerns (like rights, justice and fairness), and hiding the political as well as technical nature of these processes and decisions.

Despite platforms' constant reversion to technical solutions, 'many of these problems stem from the basic concept of filtering itself; advancements in the technology of filtering will not address all of the underlying human rights concerns' (Llansó, 2020: 2). The sociopolitical questions raised by automating content management 'will never disappear, but they may be strategically buried' (Gorwa et al, 2020: 12). And, as Gillespie asks, 'even if we could effectively automate content moderation, it is not clear that we should' (2020: 3). There is a performative role of categorizing (as 'inappropriate speech', for example), which adds to the meaning of that category. This constitutes discursive norms and understandings in the creation of social systems. It consolidates symbolic power over highly political and social issues within the private companies who design, maintain and own social media platforms.

The uses of this symbolic or discursive power can be seen in the shifting focus of, for example, regulation from trust in systems to trust in content (Mezei and Verteş-Olteanu, 2020). Metaphors link law and technology, but trust is also a metaphor, and trust in the system is an important and prevalent metaphor. In part it is an acceptance of obeying rules without knowing them, of following norms anyway based on the knowledge and expertise of others. And trust in the system is now also proxied via the metaphor of trust in governments, other users and, ultimately, the content itself. As well as addressing risk and uncertainty, trust is therefore also a sliding of epistemic power. The manipulation of trust in content enacts epistemic injustices under the power of an increasingly hidden system.

In this context, concerns like misinformation are not so much a crisis of post-truth but of post-trust. Ethan Zuckerman (2017) points out the distracting focus of the meaningless term 'fake news' and emphasizes the need to build trustworthy and strong institutions rather than fixating on specific content and the breakdown of truth. Mistrust is important in the balance of building social trust (Zuckerman, 2021), but it is often misdirected across sociotechnical assemblages. Greater focus on discursive power and the quantification of trust (and subsequent trustification of populations) enables criticism to shift from specific pieces of content designed to increase the breakdown of trust and instead bring into view the systems that enable such content and the actors who perform it.

Media platforms often operate as sociotechnical systems that not only enable but encourage the spread of these and other harmful types of content, while embedding untrustworthiness into the algorithmically mediated system itself. The Integrity Institute regularly reviews the transparency mechanism of Facebook's Widely Viewed Content Report. One such review (Allen, 2022) found that over the course of 2021, as Facebook reduced the prioritization of authoritative health information required during the pandemic, there was a decrease in top posts passing basic media literacy tests and an increase in most areas of concern. Particularly troubling was an increase in top posts by accounts that were known to violate community standards.

The review highlighted how Facebook (and Instagram) were particularly poor at ensuring the validity of top content, especially compared to other platforms like YouTube, where original content from known creators ranks highest. We could argue that this is partly due to the different cultural settings of a platform, and a different basis of trust. YouTube promotes specific creators who are often platform-specific; trust is therefore more self-contained or self-fulfilling by being based on their reputation as YouTube stars. Facebook and Instagram, by contrast, rely on a mix of interpersonal and brand reputations for trusted content. In particular, the amount of shared second-hand content creates specific issues for chains of trust across the media ecosystem including onto third-party content distributors.

The games between competing platforms have different targets of trust at different times and in different ways. Sometimes they focus on content, sometimes on advertisers.

Sometimes they exploit the assumption of a compliant userbase (though Facebook's recent decline challenges these assumptions of dominance). But all of them target policy and public discourse – from dark patterns of deceptive interface design and manipulation by algorithmic feeds to trustification as a dark pattern of public discourse.

Tech Policy Press convened a working group of academic and policy researchers on dark patterns in web/tech design ahead of a US Federal Trade Commission investigation into deceptive user interfaces. Emphasizing issues such as context, the asymmetrical impact of harms, and the systemic nature of the problem, the working group asked: 'are dark patterns the manifestation of systems, structures, code, and policies wrapped in design?' (Sinders and Hendrix, 2021). I would answer yes. The dark patterns of interfaces and algorithms not only manifest the aims of systems and policies. They are performed at the level of systems and policy, with platform CEOs regularly brought in to testify, given outsized time to justify their efforts and the legitimacy of their platform in handling public discourse and social spaces.

Meanwhile the same corporations hire policy, government and academic experts to drain opposition and gain insider knowledge to further game the systems and processes. These techniques are trustification echoed across settings. The metricized trust in content acts as a proxy for trust in systems, which is brought out in statistics to be presented to (and game the priorities or capabilities of) regulators despite the platforms themselves conducting the analysis and even framing the presentation of issues. Expertise and obfuscation return to keep alternative forms of power in the dark, a game between corporations and states competing, at base, for trust in their competence and interests from their populations.

Platforms seek to take on the role of 'trust mediators', performing the role of 'trust production' in and by technology (Bodó, 2021) – what I have here called trust extraction – despite the fact that these trust mediators are themselves largely untrustworthy. But the mediating power of platforms cares not whether it is deserving of trust. Media performs its own trustworthiness by holding discursive power and normalizing its own acceptance.

## Power and discourse

The discursive power in and of media platforms has become hegemonic in its ability to stave off criticism: questions are allowed but deflected, bringing the discourse back around to trust in innovation, expertise, and the quantifying language of power. The self-constituting discourses of technology as both means and end act to deny alternative modes of questioning. Instead, everything becomes a measurable and solvable problem. These 'problematizations' are 'the unsolved problems to which scientific articles, patents, use-cases, prototypes and proofs-of-principle propose some solution' (Mackenzie, 2015: 432), combining corporate, state and research agendas within the mediation of social problems.

By presenting problems in these specific ways, technology is presented as the solution and extracts trust through its own constitution of society as a datafied game to be completed. These aspects combine in, for example, what Roberge et al (2020) describe within the discursive power of AI narratives: 'legitimacy through performance, problematizations becoming practical solutions, and criticisms co-opted by justifications all come to inform a brave new world powered by AI' (2020: 4). They go on to write that 'if legitimacy is constructed and thus performative, it relies as well on how it is received, i.e., believed and assented to. Legitimacy, simply stated, is a symbolic give and take' (2020: 8). However, it is more often a take and take on the part of the largest platforms, establishing norms and expectations that filter down to smaller companies and shift sideways into government and academic thinking.

Tech CEOs like Mark Zuckerberg and Jeff Bezos will one moment deny they hold a monopoly and the next moment defend it. They play with the loopholes between terms and definitions to position themselves favourably in relation to media and legal frameworks. We see the overlapping and merging of state, corporate and media agendas, language and narratives. This is often backed up also by prominent researchers at elite institutions who rely on these networks for the vast sums of funding such support, such public performances of trust, can secure. Around these debates between existing forms of power, the common discourse of technology and innovation perpetuates the inequalities of race, gender, sexuality, disability and other forms of marginalization.

Counters to these discourses emphasize both social and technical approaches but must target social rather than technical issues if social justice and systemic change are the aim. For example, Aimi Hamraie and Kelly Fritsch's 'crip technoscience manifesto' (2019) seeks to resist the current ways that 'disability is cast as an object of innovation discourse, rather than as a driver of technological change' (2019: 4). Against these dominant discourses, they outline tools for 'producing new representations of disability that challenge disability technoscience discourses' (2019: 19).

This echoes M. Remi Yergeau's (2014) work critiquing tech discourse practices such as the hackathon, in terms of passing, fixing and retrofitting disabled bodies. These mainstream discourses echo the empty telethons that quantify support and action in calls and money rather than engaging with the needs of affected groups. Yergeau's response is 'criptastic hacking', which is disabled-led, emphasizing access as an ongoing action not a goal: 'We are the ones who should be hacking spaces and oppressive social systems; we should not have our bodies and our brains hacked upon by non-disabled people [...] Bodies are not for hacking. Bigotry is' (Yergeau, 2014). This type of critique aims at shifting the terms of discourse and altering power relations.

Similar approaches embodying the 'nothing about us without us' call can be seen in movements such as design justice (Costanza-Chock, 2020) that emphasize the need to centre those who are minoritized by gender (especially trans folk), race and/or class. The underlying logics are shifted here, restructuring and reordering the entire sociotechnical assemblage with a different discourse. This includes mechanisms such as a 'trusted commons' which is 'directly valuable as the basis for the existence of the social worlds of peer production, whereas it is not monetizable directly in the context of informational capitalism' (Maxigas and Latzko-Toth, 2020). It is the power structures that need 'solving', not the specific datafications of people and gamifications of social problems. Alternative, radical discourses resist the trustification of populations by generating solidarity while emphasizing heterogeneity and context-specificity, by embedding justice through relations between trust and mistrust. Fundamentally, this entails telling different stories about what technology can be and who it is for.

# SEVEN

# Case Study: COVID-19 Tracing Apps

Trust became a key feature of the way health, governance and technology discourses intertwined during the COVID-19 pandemic. The need for public trust in the measures being put in place – and the tools being designed and deployed to support those measures – was seen as paramount. So too, then, did the measurement of trust become an important factor in defining policy and negotiating the vast array of public and private organizations involved in supporting responses to the pandemic globally and locally.

This prevailing discourse focused on trust deficits, the premise that lack of trust in both government and technology was hampering society's responses to the pandemic. Trust deficit narratives combined trustification with existing deficit narratives in the perpetuation of longstanding inequalities in the effects of the pandemic. This type of framing established trust as something required of populations, and therefore something governments needed to 'get', to extract and measure, something populations were required to provide and blamed if they did not. Populations around the world therefore became subject to what I have called trustification.

The pandemic context exposed 'the complexity and fragility of trust relations at the intersection of society, technology and government' (Bodó and Janssen, 2020, 3) and 'growing societal distrust in the veracity of numbers and the increasing lack of confidence in the validity of the deployed tech solutions' (Milan, 2020: 2). But the promise of certainty offered by measurement, by quantifying social problems, became 'particularly seductive in times of global uncertainty' (Milan, 2022: 1). A reliance on data became 'a fundamental ingredient of any reporting on disease diffusion, betraying a positivist belief on the power of information to solve the most pressing problems of our times' (2022: 447–8). Trust in measurement, and measuring trust, defined many aspects of COVID-19 policy, and this is nowhere more apparent than in contact tracing apps.

Os Keyes (2020) identified in the push for technological solutions to contact tracing 'a vast array of proposals to integrate automation and technological surveillance' (2020: 59). This discourse replaced the use of human contact tracing based on consensual data sharing and building interpersonal trust developed by social researchers gathering qualitative data on transmission. These qualitative approaches were essential in informing early responses during, for example, the HIV/AIDS pandemic (Kaurin, 2020: 68). In its stead, quantitative data scraped more directly from people's devices was hailed as the answer to COVID-19, relying on technologized trust and trustified society.

A contact tracing app is a complex sociotechnical assemblage (Botero Arcila, 2020). It includes systems (infrastructure, administration, regulation), states (governments, politicians, departments and services), companies (building or running the apps, devices and/or data), researchers (designing, analysing and legitimizing approaches), the media (providing explanation and support), and the participation of other people. Many of these entities were not well trusted. Or, rather, as with the case of surveillance giants like Palantir turning their attention towards public health, they had proven themselves untrustworthy.

The issue of trust was further complicated by situations where trust in one entity relied on the trustworthiness of another. Governments became particularly beholden to the products and public image of potentially untrusted or untrustworthy private companies who could provide the technological solutions quickly enough (Bodó and Janssen, 2020: 4). Each entity also had specific groups from whom trust was rightfully absent, such as those historically excluded from healthcare systems (or receiving worse care) by race, gender, sexuality, disability or class. In this absence, trust had to be proxied and constructed.

Contact tracing apps were built on the assumption of participation. This needs either trust or compulsory enforcement to work. Many states were reluctant to make apps compulsory. But people were unlikely to trust and use an app before others had used it or it had been proven to work, or without proper independent scrutiny and assurances. In such a situation, trustification stepped in to perform the required conditions of trust, to extract legitimacy for a technological solution to a social problem.

## Debates and debacles in the UK

The divisive and difficult development of the UK's COVID-19 tracing app – NHS COVID-19 – demonstrates the problematic assumptions and structures of quantifying trust, and the trustifying processes that step in to cover over the gaps. The UK government, and the dominance of global technology discourses and corporations within the UK, had already prepared the (state, corporate, research, media) surveillance assemblages needed to enable a track and trace system. But trustification, most notably measuring trust through the proxy of adoption, came to legitimize the further use and normalization of these technologies in a public context during the pandemic. The co-opting of personal devices (in this case phones) was simply one more way that 'wearable technologies gain legitimacy and exercise authority at the level of the everyday' (Gilmore, 2021: 384). By tracing movements constantly, the NHS COVID-19 app legitimized the datafication of habits and roles at a scale and context not before deemed permissible, eliding trust to shift the norms of acceptability.

The development of the NHS COVID-19 app raised debates within research, tech and policy communities that largely came down to issues of trust – or where trust should and should not lie within the sociotechnical assemblages that would constitute such an app. Echoing the similar push in other surveillance-hungry countries such as France, the NHS started towards a centralized approach, emphasizing the need to avoid specific types of snooping attacks. The approach would work by sending Bluetooth data from users' phones to a central government database that would map contacts and send out notifications. The fear of criminal hackers was used to enforce trust in the government through political assurances of technical and operational competence at the expense of critical mistrust of that same government.

Competing methods arose with different threat models and sociotechnical priorities. Google and Apple were quick to coordinate an approach that would favour their interests and principles over the interests of governments. As creators of the two major smartphone operating systems (Android and iPhone), their dominant role assumed trust through numbers of users. They used their power over technical systems to block the UK government's centralized system, forcing a shift in government approach.

A similar perspective that eventually aligned with the Google/Apple system was put forward by a consortium of researchers: Decentralised Privacy-Preserving Proximity Tracing (DP-3T). The developers of this protocol were part of a team penning an open letter to governments, suggesting DP-3T as one alternative, signed by hundreds of academics arguing against centralized approaches. This rested on legitimate worries over government misuse of mass data collection and social mapping. But it concealed a number of pervasive assumptions: that an app-based approach would work in the first place; that this solution was clear and different enough to warrant public trust; that proximity proxied for contact (resulting in a known issue regarding vertical Bluetooth distance in denser living arrangements like cramped blocks of flats creating class-based divisions of alerts); that individuals and populations can and should be expected to trust technical systems.

Other academics in turn spoke out in favour of the centralized approach, emphasizing the external fears of hacking and decentralization. This argument relied on appeals to cryptographic validity to overcome operational vulnerabilities and government mission creep. The centralized vs decentralized debate came down to where trust should and should not be placed in the possible range of technical solutions and sources of power.

After being blocked on a technical level by Google and Apple, the UK government was forced to backtrack on centralization and adopt the tech companies' approach. This reflected the 'googlization' that Siva Vaidhyanathan (2011) had identified a decade earlier when 'public failings' led to Google stepping in where government cannot/does not address an issue. The UK government's flailing reputation of incompetence was quickly turned into an opportunity for tech company solutionism to proxy for political trust through rapid development and deployment of already normalized and socially embedded technical systems. By focusing discussion on measurable technical issues and a narrow focus on one form of privacy, 'private companies can offer means to respond to emergency time without consideration of public time' (Gilmore, 2021: 388), playing the need for rapid functionality to elide scrutiny of motives.

The UK's app continued to be plagued by issues, delays and errors at every stage, while still expecting the public to play along and participate. An Ada Lovelace Institute and Traverse report on the public view of needs for a contact tracing app (2020) not only demonstrated ways that trustworthiness could be proxied through the perceived academic or democratic validity of independent audit by researchers, but also showed how trust in apps was influenced by trust in other government actions. Despite this need to build and demonstrate trustworthiness, numbers again were used to justify the app. For example, the announcements of the launch of the app highlighted the number of businesses that had already downloaded a QR code for visitors to scan through the app to log their location for contacts (Downey, 2020).

Continued reporting of the app focused on the number of downloads and user accounts as a measure of success through uptake. But how much uptake is enough? Enough of whom? The Ada Lovelace report highlighted how one in five in the UK do not have a smartphone, creating further inequalities in access to healthcare and public health protections even if the app had been successful.

The focus on percentage uptake was rooted in the prestige and epistemic credibility of Oxford University, who released a report claiming that 'the epidemic can be suppressed with 80% of all smartphone users using the app, or 56% of the population overall' (Hinch et al, 2020: 3). However, this was based on certain assumptions, relying on mathematical (speculative) modelling, and later justified through further modelling and prediction. This repeated the quantification compulsion to reduce complex social issues to computable numbers that perform their own validity. By focusing on quantifiable proxies such as device and user numbers, the massive inequalities in the UK and elsewhere were pushed under the rug.

A combination of cryptographic and epidemiological logics combined to instil trust through what became a key number of 60 per cent adoption for tracing apps. The media was quick to jump on this clear and easy to understand metric, spreading the assumptions of the Oxford study not only across the UK but globally as well. The reduction of complex issues to a single measure of adoption created a single numerical target that collapsed the context specificity of a whole range of existing and new social inequalities that deserved greater attention.

The epistemic dominance of trust in Oxford's credibility overrode many critical voices highlighting the potential harms or inequalities in the rollout of an app, but also further entrenched the assumption that a quantitative app–based approach was the right one for the whole world. The need for rapid responses and the huge resources that enabled such quick analysis by a dominant academic institution furthered the colonial practices of epistemic injustice that imposed technological solutionism on global responses to the COVID-19 pandemic.

## Global perspectives

The existing resources that wealthy nations and their tech corporations could throw at the issue of tracing the spread of the pandemic enabled a setting of political, epistemic and discursive framings that privileged quantification and solutionism. This could be seen in the spread of contact tracing apps across the world. The global adoption of a specific tool erased the specific needs of local contexts. But the different infrastructures, regulations and rights, levels of access and participation, and cultural settings within these contexts produced different effects and harms in different locations.

India, where the persistence of colonialism would lead to the production of vaccines inaccessible to its own population, is one example of intertwining global narratives placing trust in an app-based solution amid an existing problematic relation of trust with state–corporate data systems. This mistrust was rooted in the highly problematic Aadhaar biometric system that paved the way for further problems of state–corporate privacy violations and unjust design practices (Raghunath, 2021: 201). Such issues were repeated within the Aarogya Setu COVID-19 tracing app in its reliance on identity markers. This was particularly dangerous within the context of right-wing groups blaming the Muslim population for spreading the virus; migrant populations and other 'unentitled' groups (Milan et al, 2021: 18) were particularly harmed by the links between Aarogya Setu and Aadhaar, which exacerbated the exclusion of those not 'counted'.

This challenges the Anglo-European narrow focus on privacy over wider structural issues surrounding data, state and corporations, particularly in the context where the app was released without the safeguards of robust data protection laws (Das, 2021). The Aarogya Setu app itself had one of the fastest download rates of any app, leading to it being hailed a huge success and suggesting that, despite issues, there was trust in this technological response. However, while the number of downloads was high, it only represented about 7 per cent of India's population. By altering the measurement and excluding the context, the government gamified the population to achieve its goal of generating the conditions of trust despite a lack of trustworthiness.

India has low rates of smartphone ownership. When the app became mandatory for various groups (including those who had broken curfew) or activities (such as entry into certain venues or freer movement as lockdowns eased), large groups were excluded. The blinkered quantification of success was used to promote trust in an invasive tool. This tool not only exacerbated state surveillance but created multiple dimensions of exclusion for those forced to download the app and those unable to benefit from the increased movement it allowed.

Mandatory apps were also imposed in Qatar and Bahrain. Compulsory adoption performatively and forcefully constitutes the conditions of trust in the government and technology without the need for trust. This was despite Amnesty International and other rights groups highlighting known security flaws in Qatar's app (2020a) and labelling Bahrain's app one of the most invasive ones deployed (2020b).

Similar issues of control and trade-offs were seen in South Korea and China. Yeran Kim (2021) highlights the biopolitics involved in the rise of surveillance-driven AI as a response to the pandemic in South Korea. The discourse of maintaining 'normal life' was mobilized as a trade-off for mainstream discourses that attempt to 'naturalize' technologically mediated assemblages of governmentality, including additional impositions on marginalized groups such as non-nationals. This speaks to wider uses of trustification in AI, data and technology discourses more generally. Both South Korea and China used campaigns based on 'finessing the blurred boundaries between care and control' (Kim et al, 2021). This enabled a framing of paternalism rather than aggression to perform the conditions of care as a proxy for trust to legitimize the expansion of state surveillance regimes.

Other countries used the familiar false trade-off of security as a proxy for trust. For example, Indonesia and the Philippines leant on security-heavy approaches to overcome weak healthcare systems, including the leads of their COVID-19 responses being military rather than public health officials, creating low trust in tracing apps (Poetranto and Lau, 2021: 99). The Philippines also made efforts to silence post-pandemic data custody issues (Lucero, 2020).

The Argentinian government focused on tracking individuals. The CuidAr app for self-diagnosis and symptom reporting, mandatory for those travelling from abroad, was pushed under the narrative of smart lockdown easing, creeping the functionality and value of the app to perform the conditions of enforced trust. Meanwhile the CoTrack app that enabled additional geo-tracking features for contact tracing was developed separately from government but was adopted by some provinces seeking the enhanced capabilities of private technological solutions (Álvarez Ugarte, 2020: 87).

Mexico simply bypassed the need to rely on tracing apps, combining anti-terrorism tools from NSO Group (makers of the Pegasus spyware undergoing multiple legal challenges around the world) with mobile operators' mass data (Cruz-Santiago, 2020). However, demonstrating the differences of context, in Israel the higher level of trust in the security services (Shin Bet) enabled a repurposing of anti-terrorism tools for COVID-19 tracing, even though the Shin Bet themselves resisted turning their tools to their own population (Gekker and

Ben-David, 2021: 149). Here the trust relations were situated in the normalized surveillance practices of marginalized and external groups.

Elsewhere, the extraction and imposition of trust in centralized government solutionist narratives generated additional epistemic violence in prioritizing quantification over context and alternative forms of knowledge. The design of the Aotearoa (New Zealand) tracing app involved 'little meaningful engagement with principles of Māori Data Sovereignty' (Cormack and Kukutai, 2021: 142–3). This worsened existing problems and created further issues of trust, including the continued epistemic violence of excluding Māori expertise, perspective and context over data rights. As Cormack and Kukutai write, 'Māori knowers and knowledges have been marginalized, and unjust data practices continue to privilege the priorities of the dominant Pākehā (NZ European) population and wilfully ignore Māori data rights' (2021: 143).

On a global scale, the proxy of adoption for trust joined with a focus on Anglo-European nations. For example, one 'cross-country survey' (Altmann et al, 2020) used acceptability as a proxy for trust. It also included only the US and four European countries in its analysis. It may have been technically cross-country but it hardly offered a global perspective.

Even within wealthier nations with established technological and political infrastructure, 'digital contact tracing will exclude the poor, children, and myriad other uncounted groups' (Alkhatib, 2020). These exclusions included those without devices or who share devices within a family, those in settings where major spreading incidences might occur (prison, care, education, food production), and those whose movement is not voluntary (Gupte and Mitlin, 2021). They were rooted in assumptions of class, race, gender and power.

This marginalization (re)produced within the pandemic a '"standard human" based on a partial and exclusive vision of society and its components, which tends to overlook alterity and inequality' (Milan, 2020: 1). Beyond thresholds of trust or acceptance, this led to the continued uneven distribution of whether technological solutions 'work' (D'Ignazio and Klein, 2020b). The epistemic framing of the pandemic entrenched issues of who 'counts' in quantifying social issues, often excluding based on gender, class, race and other socially constructed categories of marginalization.

The extractive exploitation of the Global South echoed the internal exploitation of marginalized key workers within the Global North. We can see networks of extraction across the whole sociotechnical assemblage of contact tracing apps. From the extraction of materials to the manufacture of devices to the design of systems that ignored compulsory movement for low paid but essential jobs, 'the dynamics of racial capitalism necessitate a multifaceted understanding of extraction, exploitation and oppression' (Aouragh et al., 2020). Extraction and dispossession were stabilized and concealed through global historical infrastructures, while privileged populations were trustified into adopting technologies that exacerbated these intersectional injustices not only within the frameworks of datafication but also within the effects of unequal public health policies.

Datafication supported trustification throughout the pandemic. COVID-19 tracing apps demonstrate the different mechanisms through which states, corporations, researchers and the media perform and extract the conditions of trust from populations to legitimize the use of surveillance apps on populations. But the harms (and trust requirements) of these apps were not felt evenly. They echoed wider inequalities whereby already marginalized groups are exposed to the risks and expected to act as if they trust technologies and those designing and using them, while these impositions are used to game trust in the system from those already privileged. It is important to highlight the multiple dimensions of marginalization in local and global contexts, in specific apps and in the app-based discourse more widely. Otherwise, the only survivor of global crises will be the power of technology discourses.

# EIGHT

# Case Study: Tech for Good

Trustification operates when populations or groups are forced to perform the conditions of trust without corresponding relations of trust. The power asymmetry between those extracting trust to legitimize technology development and deployment, and those expected to trust them, creates and entrenches structural inequalities. This occurs even within initiatives that claim to be creating technologies 'for good'.

These programmes – data for good, AI for good, tech for good – seem to be everywhere, connecting corporate, state, research and public sector organizations, agendas and interests. But we must ask what good? And whose good? The interests of those on whom these systems operate are often left out of the decision-making process, sidelined in the discourses that determine but rarely define this elusive 'good'. And tech 'for good', especially humanitarian tech, is used to extract legitimacy for tech 'for no good', furthering capitalist and/or state aims of escalating and consolidating power and wealth.

The discursive project of tech for good performs many of the dominant (and harmful) assumptions that surround the use and role of technology in society. The quantification of social issues, and the datafication of those affected by them, feeds into the expansion of technological solutionism and the extraction of legitimacy for those solutions. Involvement in humanitarian projects therefore acts as marketing for tech companies, what has been labelled as 'aidwashing' (Martin, 2023) in the embedding of surveillance systems to control the distribution of aid, or #Help (Johns, 2023) in the policy influencing practices of translating populations to data, embedding the interface as a mode of control in ways that echo the dashboard of sensory power (Isin and Ruppert, 2020). This is all framed in the prevailing narratives that label tech as neutral, tech companies as benevolent, and their solutions as objective.

Around technologies such as AI, the assumption that it 'is neutral and should be used "for the greater good" […] neutralises criticism as simply a matter of imperfect information.' (Jansen and Cath, 2021: 189). These 'completionist' desires become embedded in sociotechnical systems. The framing assumes that tech for good is a universally viable approach and aim, and that any given technological solution can be applied to any context with enough data and infrastructure. These assumptions generate the justification for increased surveillance practices, embedding technological power in ever more places.

No matter the problem, there is a tech company ready to save the day. It seems that any humanitarian crisis now receives a flood of (often empty) promises of support and generic solutions from tech CEOs, backed by governments giving them seats at high level tables where responses and policies are determined. But approaches and projects like 'AI for good' are 'an "enchantment of technology" that reworks the colonial legacies of humanitarianism whilst also occluding the power dynamics at play' (Madianou, 2021: 850). Humanitarianism is often dehumanizing. It is often fraught with solutionist assumptions and the perpetuation of injustices while justifying continued power asymmetries. Saviour complex spreads from wealthy powerful governments to wealthy powerful tech companies, wrapped up in the 'enchantment' that tech will save everyone.

A critical response is therefore to confront the assumed conditions of trust that conceal the imposition of colonial and capitalist power. Political trust can be linked with political disenchantment, and thereby alternative modes of participation (Eder et al, 2014). Reinserting mistrust in political institutions (Zuckerman, 2021) can offer transformative power. There needs to be room at the table of global humanitarian responses, dominated by the Global North, for those most affected by the inequalities generated by crises amid unequal social structures. Giving voice to mistrust can push towards alternative responses that embrace locality and plurality rather than imposing one particular view of what a universal 'good' looks like.

## What good?

A key challenge to the legitimacy of tech for good is to ask how good it really is. Discourses of technological solutionism are everywhere but they often leave the effectiveness and relevance of that solution to the imagination. Tech for good always risks being a solution looking for a problem. It is this framing that makes apparent the use of tech for good as a legitimizing factor for the same technologies elsewhere.

Humanitarian projects create a reputational boost for companies seeking to avoid an image of surveillant practices or biased automated decision making, in the same way that funding charity and humanitarian efforts boosts the social capital of billionaires to distract from the extractive ways they have amassed their great wealth. These often overlap, like the Gates Foundation playing an enormous role in influencing research agendas 'for good' as well as the areas to which they are practically applied. Other tech CEOs have followed suit. The Chan Zuckerberg Initiative and the Bezos Earth Fund seek to mitigate the reputational harm of their companies' societal and environmental harms, while entrenching specific (often US, capitalist, techno-solutionist) values in global humanitarian, regulatory, research and social infrastructures.

Beneath these media-friendly aims lie claims of solutions that are often smoke and mirrors. As Frederike Kaltheuner writes, 'when it comes to predicting any social outcome, using AI is fundamentally dubious' (2021, 9). The 'fallacy of AI functionality' (Raji et al, 2022) not only perpetuates a false enchantment of what technology can actually achieve in complex or unstable social contexts, but it also shapes how society responds.

The assumption that technology 'works' makes it particularly difficult in international humanitarian applications. Humanitarian tech often spans regulatory systems in ways that extract legitimacy through the relative positioning of organizations, companies and the groups they are working with (or, more often, working on). What recourse is there for an affected group when they are prevented from accessing or understanding in-depth information about how a sprawling AI and data system works, when the policy makers who allow it likely do not know themselves? With the increase in private–humanitarian partnerships, verifying trustworthiness becomes a high stakes game for those seeking to 'do good' without simply acting as a mouthpiece for the interests of tech companies.

For example, the UN AI For Good initiative has significant investment and influence from private companies (as well as various governments). This seeps into the framing of problems with language like 'acceleration' and 'innovation', making connections between 'innovators' and 'problem owners' or the United Nations Sustainable Goals. Its programme of speakers and events is focused on innovation (and innovators) rather than affected communities.

Similarly, the UN Sustainable Goals 'Act Now' app and chatbot implores people to log individual climate actions, feeding into the datafying and gamifying logics of corporate platforms and the app-for-everything solutionist mentality to proxy for systemic and structural social and political change. More productive (but more hidden) are some of the actual resources on the app such as how to speak up for climate action to governments/companies, or chatbot support for humanitarian organizations, palming them off on an automated response.

Chatbots have a questionable validity. They are often used to justify the expansion of AI, with little evidence of the usefulness of such resource-intensive technologies. As Mirca Madianou

(2021) explains, chatbots instrumentalize communication, creating 'disconnects between affected communities and aid agencies' (2021: 864) and 'prioritise information dissemination, a process which doesn't require the complex infrastructure of AI' (2021: 858). The technologies are solutions looking for problems; there is often little need for that level of invasion, automation or investment. The information that is disseminated most often supports the prevailing discourse that giving financial and decision-making resources to tech companies is the answer rather than providing affected people with the resources and agency they need.

The labelling of technologies, or particularly of data, as 'humanitarian' carries with it a discursive performance of purpose, virtue and usefulness. For example, the Humanitarian Data platform maintained by the UN Office for the Coordination of Humanitarian Affairs takes a specific definition of humanitarian data that reduces context to quantifiable units like population stats and geospatial units, occluding the significant role of sociopolitical and (infra)structural issues.

Standardizing formats constitutes what 'counts' as 'data', and by extension who counts in data or as a controller of data. It enforces norms through metadata, with differences between private, public and requestable datasets: private datasets are not required to provide a methodology, for example. It also performs meaning through symbols. A green leaf stands for up-to-date data, for example. This establishes a specific narrative prioritizing the 'health' of data, far removed from the sustainability or environmental impact a green leaf might normally imply (especially in humanitarian contexts). All these aspects performatively shape the rules and norms of what data can be shared. This creates additional discursive power by defining what can be said, how it can be said, and who can say it.

After various UN agencies and consortia of NGOs, the biggest corporate contributor of datasets to the Humanitarian Data platform is Kontur. This company markets itself as a 'disaster ninja' to advertise their 'solutions' to all known problems by collecting all available data, a terrifyingly dystopian set of aims placed in full public view. They are followed on the list by Facebook/Meta, showing the further platformization of the humanitarian sector.

The Humanitarian Data project's associated 'State of Humanitarian Data' report (OCHA, 2022) presents pages and pages of tables showing estimations of percentage completeness of 'relevant' data across types and countries. This shows a record of the spread of data colonialism and completionist quantification narratives that seek to justify reducing everything to data that can easily be shared between governments and corporate entities. The Humanitarian Data platform also offers users the ability to 'follow' data as if it were a friend or a trend, proxying trust in organizations with trust in datasets. This not only legitimizes and quantifies human suffering but gives it celebrity or influencer status.

The sliding of proxies for trust and other social issues inserts the datafying, quantifying narrative as a flaw in many tech for good projects. Turkey's Data4Refugees project in collaboration with mobile phone networks entailed mass collection of location and call ID data with a high level of granularity. Labelling callers as 'refugees' or 'citizens' was supposed to assess integration over time. But not only does this create a permanent rift between the categories of refugee and citizen, it mediates the persistence of administrative barriers and surveillant infrastructures. The evaluation of integration relied on several proxies, things like calls between refugees and citizens as indicators of integration.

The project acknowledged certain problems, like the compulsory nature of calls to citizens (especially, for example, government employees) to access services, or offline communication within and between refugee groups within specific areas. But these limitations did not stop the initiative continuing to use these problematic proxies or affect the way they presented their results.

The challenge call for projects further demonstrated the solution in search of an answer. With data that was largely collected anyway, the humanitarian angle emerged as a justification for national surveillant networks and power asymmetries. It feeds into the extractive logics of 'collect data first, ask questions later'. This shows the way that legitimacy is not required from those on whom these processes are inflicted. Instead, a humanitarian aim is used to extract legitimacy from the mainstream citizen population at the expense of refugees, and no doubt to normalize the use of these surveillant systems to later turn them on those same citizens.

This assumption that it is simply fine for companies to collect vast amounts of data if they can legitimize their practices by also making it available for humanitarian purposes is embodied in Facebook/Meta's 'Data for Good' project. Existing data about Facebook connectivity is used as a proxy for a number of measures: population; demographics; movement (especially in crisis); wealth; and social connection between places. These data points make assumptions not only about whether Facebook connectivity can stand in for various issues but also that such proxies are evenly distributed enough to be viable.

This data is also connected with satellite imagery and AI to supplement measures such as population density and wealth, quantifying further social judgements. These measures are based on a supposed 'ground-truth' of surveys conducted in the Global South by the US Agency for International Development. This is an organization open about its furthering of US interests within its humanitarian work, highlighting the state–corporate power at work in the continued intertwining of colonialism and capitalism.

The development of these large datasets shows the connected data ecosystem already in place and the power Facebook already wields. It assumes Facebook as infrastructure and as proxy measure for a range of social concerns, an embedded power that influences how governments, researchers and humanitarian organizations frame their assumptions. There is often no choice but to use their data during crisis and humanitarian responses.

The ways Facebook constitutes the conditions of trust by holding power over data, connectivity and social infrastructure emphasize the rapid deployment or repurposing enabled by innovation narratives. Facebook (and Meta) are always there, always ready, with data already in hand. Facebook performs trustworthiness to legitimize constant and expanding data collection and development of analytical tools, further trustifying corporate technology and quantification logics within government, research and humanitarian contexts.

## Whose good?

The agendas of power are not homogenous. The competing interests and discourses of state, corporate and research logics play a game at the level of population measured most often in projects, datasets or funding. These forms of power are often united in their colonial/capitalist underpinnings, but manifest with different emphases and different loci of epistemic and material power, framing debates and reaping the benefits of the solutions employed. Similar narratives support these different interests, tales of objectivity, reductionism and completionism. We see echoes of the historical European 'intellectual necessity of the idea of totality, especially in relation to social reality' (Quijano, 2007: 174).

From these underpinnings, 'innovation' spreads across sectors, for 'to align yourself with "innovative" narratives gives you a head over the competition' (Milne, 2021: 119). An area such as humanitarian work should not be a matter of competition. But at the level of discourse the games of power continue, and inappropriate technological logics become embedded all the way up to the UN. Meanwhile, 'lack of care characterizes a lot of the problems around hype in science and technology' (Milne, 2021: 122), and care ethics or a duty of care become lost under the quantification of power with populations (number of people helped; amount of public funds allocated) as their resource.

For example, the UNICEF Magic Box is a collaborative platform developed in conjunction with researchers and governments, but is also heavily reliant on private companies. The initiative manages investment in 'frontier tech' linked to UNICEF's Innovation Fund for technology 'solutions'. This includes the UN CryptoFund as well as leads for AI, data, blockchain and drones. Innovation and capitalist-colonial data and technology narratives prevail, leading to an emphasis on business growth and product development using funds aimed at helping children.

Trustworthiness in this setting is constituted through business hype and marketing, like the founders and other staff shown with pictures of them as children, echoing 'quirky' corporate branding within humanitarian contexts. The 'buzz' of innovation embeds solutionism, decontextualization and depoliticization while dispersing and proxying trust across vast sociotechnical assemblages. This sinking of specific issues and actions into the wider milieu of technology development generates 'an imagined absolution of responsibility, a false narrative in which they've created an artificial system outside of anyone's control, while the human population affected by their decisions and mistakes is inappropriately erased' (Raji, 2021: 60). The sprawling nature of both technologies and humanitarianism leave tech for good initiatives to fall through multiple cracks of accountability, providing further opportunities for avoiding scrutiny and amassing colonial power. And they, of course, gain wealth and reputation along the way.

This power is based on a 'fundamental asymmetry [...] at the heart of every humanitarian intervention including those involving "AI for good"', an asymmetry in 'the power relations between mostly western "saviours" and the suffering "others" who are typically former colonial subjects' (Madianou, 2021: 852). This power structure legitimizes material, symbolic and epistemic violence on those it is claiming to 'save'. This last point, epistemic violence, is evident in the power of tech for good – particularly data for good or AI for good – to shape how issues are framed (as quantifiable problems requiring technological solutions), based on the legacies of colonial epistemologies that position Europe as a legitimate focus of power to justify extraction of knowledge, resources and people.

Against this 'arrogant ignorance' (Icaza and Vasquez, 2018: 112) common to academic, state and corporate logics, decoloniality focuses on confronting the claims of 'objective, zero-point epistemology from which universal assertions can be made and thus other knowledges and ways of being eradicated' (Mumford, 2021: 4). Alternative approaches – such as combining Ubuntu philosophy with human rights in AI governance (Mhlambi, 2020; Birhane, 2021a) – emphasize different epistemic frameworks relevant to the specific contexts in which they are deployed and encourage pluralism in wider developments. The aim here is not a different form of measurement, not a new form of enclosure, but an expansion of relational approaches to tackling issues of systemic injustice and alternative practices of designing and using technology.

However, this cannot be collapsed into a single issue such as geographic location. It is not merely enough to develop technology 'locally'. For example, the company Analytics Intelligence seeks to make data and algorithmic technologies 'accessible to all'. Yet its aim to become 'the Palantir of Africa' (Onokwue, 2020) replicates colonial powers. When this includes tools such as facial recognition for immigration management, other forms of marginalization must be considered.

We need to understand 'the margin' (Rodríguez, 2017) here as 'a shortcut to speak of [...] processes of asymmetrical access to material and symbolic resources shape differentiated and unequal access to the public sphere' (2017: 56). This can also be thought of spatially as 'complex sites of "otherness", as well as inequality and power struggles' (Masiero et al, 2021: 17).

Across the sociotechnical assemblages of tech for good are perpetuated not only the extractive desires of colonialism and capitalism but also the asymmetries of class, gender, race, disability, age and many other loci of injustice. It is this diversity of situations that needs highlighting, the 'urgency of addressing the data-centric regime from a perspective that considers their growing complexity, dynamism, and the relevance of situated contexts' (Ricaurte, 2019: 356). Similarly, diversity must be accompanied with radical challenges to the power structures that frame and support injustice in a multitude of contexts.

The discursive landscape of tech for good extracts trust through the assumption that humanitarian interventions always constitute this 'good'. The performed legitimacy of this flawed assumption perpetuates injustices through proxy chains of trust and an avoidance of scrutiny and accountability. Even promises of participation and accountability are often 'ultimately only offered [as] a box-ticking exercise' (Madianou, 2021: 860) in which legitimacy is extracted from those upon whom it is imposed by generating appearances of accountability through asymmetric communication and power structures (Madianou et al, 2016). It is here that the push – across state, corporate and research sectors – towards efficiency and innovation reveals extractive and oppressive logics that reduce affected people (often in the Global South) to a resource used to extract legitimacy from privileged populations (often in the Global North). This discourse is normalized to support unequal global power structures in a wider game among those already holding power. Tech for good demonstrates the cascading forms of trustification that entrench technology narratives at the expense of context and people.

# NINE

# Case Study: Trusting Faces

The face, as a site of interpersonal interaction often linked to identity, has become a prime target for quantification, datafication and automated assessment by algorithms like AI. Facial recognition has seen a massive rise in use across high stakes contexts like policing and education. The systems are particularly used to assess trust and its various proxies. These technologies are an update of historical practices like physiognomy that assign values for personal qualities based on the shape and measurements of people's faces (or brains or other body parts).

Parallels between AI and physiognomy are 'both disturbing and instructive' (Stark and Hutson, 2022: 954). Both are an attempt to read into the body characteristics that are deemed innate, and therefore inescapable. Measuring for these attributes performs them as roles and identities for specific individuals and groups. They include characteristics like 'employability, educability, attentiveness, criminality, emotional stability, sexuality, compatibility, and trustworthiness' (2022: 954). Trustworthiness was a key target for physiognomy, attempting to embed social biases within the narratives of a scientific discipline, performing legitimacy for discrimination. Parallels also include the view of physiognomy as progressive, which aligns with innovation narratives that hype up facial recognition 'solutions'.

However, despite the sheer number of tools and academic papers being developed, there are big questions over the significant flaws in methodology and assumptions. This includes a frequent lack of definition of what concepts like trustworthiness actually are, part of a chain of proxy conflation of terms and ideas (Spanton and Guest, 2022). This critique highlights the ethical risks inherent even to 'demonstration' type academic papers, aligning with what Abeba Birhane describes as 'cheap AI' rooted in 'prejudiced assumptions masquerading as objective enquiry' (2021b: 45). Faulty demonstrations of the possibilities of machine learning technology are easily picked up and misinterpreted by the media to feed the myths of AI. But problematic papers also enable the same biased and decontextualized judgements to be easily transferred to policing, insurance, education and other areas where quantifying trust is directly, materially harmful.

## Saving faces

The way facial recognition technologies are developed and deployed perpetuates specific injustices. If they work, they purposefully entrench power. When they do not work, which is often, then 'the harm inflicted by these products is a direct consequence of shoddy craftsmanship, unacknowledged technical limits, and poor design' (Raji, 2021: 57). What questions do these systems even ask? What questions do they answer? The two seldom align. These sorting technologies become a technological solution looking for a social problem. If they cannot find the question they want to ask/answer (or cannot voice it, if it is one that would be explicitly racist or otherwise discriminatory), they use the same result to answer a different question, creating and legitimizing proxies for inequality and injustice.

Whether they work or not, these systems – taking systems in their broader social sense – cause significant and multiple harms. And, because 'for any individual group or situation, classifications and standards give advantage or they give suffering' (Bowker and Star, 2000: 6), it is not about a linear measure of harms versus benefits. It is about who is harmed and who benefits at their expense, and how such technologies are therefore 'cheap' for those producing and using them but have a high cost for those harmed (Birhane, 2021b: 44). Simone Browne (2015), Ruha Benjamin (2019) and Safiya Noble (2018), among others, have charted the long history and contemporary impact of surveillance and algorithmic sorting based on the colour of people's faces and the attribution of race (and racial hierarchies) to those categories.

These forms of data power can be seen in CLIP, a visual semantic model developed using a vast dataset of images with annotated labels about their content, used for a range of computer vision tasks including identifying people according to various categories. For example, there are issues with the generic 'person' label and who counts as a person compared to the imposition of more specific (and potentially discriminatory) labels (Wolfe and Caliskan, 2022). Significantly more White faces were given the label 'person' while other skin tones categorized by racialized descriptors; men were also more likely to be labelled 'person' than women were, while women over 40 were more likely to be defined by their age.

All this adds up to the way the default person is constituted as White in CLIP, while it also shows 'hypodescent' of biracial individuals into non-White categories, more pronounced among women (Wolfe et al, 2022), further automating exclusionary practices and embedding racial hierarchies within the dataset. These sorts of issues with CLIP, as with many other models and datasets, are partly due to using automated scraping/crawling to compile data. By contrast, more active curating could enable the embedding of more socially aware and just language (Birhane et al, 2021). Trust in automation performatively constitutes the entrenching of extreme bias as the norms of technical systems; the false narrative of objectivity standing in for confronting the complex social issues at stake. The appearance of neutrality performs bias as merely reflecting rather than also producing inequality.

Alex Hanna, Emily Denton and others use critical race methodologies to highlight the flawed construction of 'race' in facial recognition systems. This is based in the problem that 'Race is a major axis around which algorithmic allocation of resources and representation is bound.' (Hanna et al, 2020: 501), and yet the socially constructed nature of race is rarely considered. This leads to racial categories being performed as objective characteristics rather than social constructs, which in turn perpetuates the discourses of technological neutrality and the eliding of structural injustice. This is a political issue that requires a political and social understanding, and political and social responses rooted in historical mistrust and a realignment of power.

Extending Wendy Roth's separation of different dimensions of race (2016), Hanna et al propose more specific uses for different conceptualizations around social and physiological aspects that are currently proxied for one another under the label of race. For example, self-identification of racial identity is different from racial self-classification to others. This separates inwards feelings and lived experience of race from external categorization within specific contexts, while both raise concerns when interacting with administrative systems containing a limited set of options. Observed race by others is further separated into appearance-based and interaction-based, separating the cultural markers attributed to (for example) faces from behaviours and relationships. Reflected race is how we think others view us, internalizing specific and cultural expectations and, finally, phenotype covers the more physiological features such as skin and hair colour themselves.

Different dimensions have different uses. Self-identification might be useful for looking at political beliefs and attitudes, whereas observed race is helpful when thinking about various forms of discrimination in different social contexts, while phenotype might be useful alongside self-classification in assessing unequal health outcomes. And we can also think across Hanna et al's categories, for the different aspects of any complex social issue that require thinking from multiple perspectives, multiple narratives, and confronting the intersecting matrices of oppression.

Race is not, then, what is really under scrutiny in facial recognition, and we should ask whether race (or any other category) is even a relevant factor to consider in the design of such systems (Hanna et al, 2020: 510). Race comes into question when we start to unpick the different ways that faces are categorized and inequitably operationalized against people. This is not a matter of better, more specific, categories, but about more appropriate, contextual and flexible *types* of categories: dimensions of race, which are themselves unstable. It is about 'who has the power to classify, to determine the repercussions/policies associated thereof and their relation to historical and accumulated injustice' (Abdurahman, 2019). It is important to uncover who controls the discursive framing of race and other descriptors of people's faces, bodies, identities and lived experiences, especially when these are embedded in technical systems and sociotechnical assemblages that are used to make decisions affecting people's lives.

## Performing categories

Technologies of classification make social categories material, naturalizing them and providing an appearance of objectivity. Quantification attempts to legitimize their use in marginalization. The face has become 'a particularly productive site for such projects, with its classification tied to determinations of personality, ability, and morality in ways that justify hierarchies of race and gender/sex' (Scheuerman et al, 2021: 2). The history of these technologies lies in colonial projects not only of race but of gender, including disciplining those who do not fit the dominant binary discourse. Colonialism and the assertion of hierarchies are echoed in the way 'contemporary auto-essentialization of gender via the face is both racialized and trans-exclusive: it asserts a fixed gender binary and it elevates the White face as the ultimate model of gender difference' (Scheuerman et al, 2021: 1). It shows how simply adding another category of trans or non-binary, for example, does not undo the erasure of non-Western gender/ sexuality practices that predate and persist despite the epistemic conquest of apparently 'objective' colonial discourses.

Platforms and the design of data-driven systems often operate a gatekeeping of categories in the use of automated gender recognition. Giggle, for example, is designed as a women-only space using facial recognition – specifically automated gender recognition (AGR) – to verify access for one specific gender. Non-binary and transgender people tend to confound AGR systems, which label them as othered, as sexualized and as a threat. By performing gender binaries as a basis for constraining access, Giggle's use of AGR attempts to constitute 'a female essence, one that is discoverable via the face and which rises above the confounding effects of "diversity" (which they claim technology can account for)' (Scheuerman et al, 2021: 10). This integrating of quantified faces linked to trust and access legitimizes colonial and transphobic discourses, and 'auto-essentialist tools like automated facial analysis must be accountable to the histories that have shaped them' (2021: 12). Yet they also introduce a new set of technological justifications as solutions for problems of identification, all under the computational logics of quantification and the assumptions that such 'recognition' is viable and justified.

As with race, gender becomes a proxy for many other things, like 'the more precise values that inform purchasing decisions' (Keyes, 2018: 15). The collapsing of people into specific, limited, fixed categories introduces particular issues around trans inclusion. A study on misgendering in automated systems found that 94.8 per cent of HCI papers using AGR saw gender as binary, 72.4 per cent as immutable, and 60.3 per cent as physiological. This reflects deeply entrenched practices of 'a remarkably consistent operationalization of gender' that erases the possibility of trans and other genders through the discursive and categorical framing of data (Keyes, 2018: 7). AGR (like other forms of facial recognition) captures only quantifiable physiological elements, not social ones. This legitimizes the way AGR projects treat non-binary, mutable and/or social expressions of gender as a threat.

Despite these technical, discursive and social limitations, AGR is being embedded in broader systems of categorization, and misgendering (particularly of trans and non-binary folk) is becoming embedded in such systems (Hamidi et al, 2018). The symbolic violence of misgendering in these harmful technologies includes deadnaming (using a trans person's prior identity), outing (revealing someone's gender or sexuality status without their consent), as well as attributional harms of using the incorrect gender (and basing decisions on this misrecognition). Automated gender recognition reduces people's complex identities and experiences (which may shift over time) into a specific number of rigid categories.

The performative construction of gender binaries (Butler, 1990) is entrenched into contexts and norms that further impose potentially harmful gendered expectations in the design and application of new online social and administrative systems. Misgendering has been described by study participants as 'the worst social exclusion' that is 'just going to exacerbate what's already there' (Hamidi et al, 2018: 5–6), shifting from individual human bias and exclusion into additional forms of sustained systemic harms. The problem highlights the fallacy of sorting people in this way, the inherent flaws in categorizing and quantifying aspects of human life that are, and should remain, beyond categorization.

Joy Buolamwini and Timnit Gebru (2018) demonstrate the need for intersectional critique, as the error rates of major facial recognition systems were not only higher on women than men, and on darker skin tones than lighter ones, but darker skinned women had an error rate far higher than simply combining those two categories. In framing their 'gender shades' study, they made particular note of the unstable nature of race and ethnicity labels – demographic categories used that imply social rather than physiological features; those categories may vary geographically and over time; and measures like skin tone have huge variation within any given category (2018: 81).

In the context of Hanna et al's (2020: 510) critical race framing, however, this study acknowledges but still risks perpetuating harmful binaries of both gender and race. This is explained within the study as being due to the fact that the original datasets in question were labelled according to these categories. While the work had a demonstrable impact in improving the accuracy of facial recognition systems (Raji and Buolamwini, 2019), a limitation of audit and dataset benchmarking approaches is that it requires engaging within the terms set by the oppressive systems themselves. The wider intersectional critique in which Buolamwini and Gebru's paper and activism (through groups like the Algorithmic Justice League) are situated also looks at breaking down multiple social categories, highlighting the narrative, mutable aspects of identities and lived experiences that elude fields on a dataset.

Beyond targeting flaws in datasets, we should also ask whether the intended function and embedded priorities of an algorithm are trustworthy, and who stands to benefit or be harmed. There is a broader social need to reclaim the framing of the question around whether faces should even be used for classifying people in the first place. The face might be the first thing we see of people, but that does not mean it should be considered an automatic site of datafication to assess the trustworthiness of individuals while imposing trust in those flawed systems of judgement.

## Getting emotional

If there are huge problems with machines attempting to quantify and categorize the characteristics of faces, there are even bigger problems when they try to read expressions on those faces. A study into non-expert perceptions of AI facial inferences and the justifications for them in different contexts (Engelmann et al, 2022) found that AI was deemed not able to judge 'second-order' inferences like trustworthiness, whereas 'first-order' inferences (like skin tone, emotional expression and to a lesser extent gender) were largely accepted in low stakes settings (like advertising) but not when stakes were higher (like job interviews). Where the success and relevance of the technology was used as the primary factor in its appropriateness, where inferences were deemed successful, then other issues such as harms and bias were used to justify why they should nevertheless not be used. Some questions remain around challenges to the categorization of gender and emotion, and how skin tone links to racial sorting, but the study does demonstrate a wider rejection of facial inferences with contextually specific justifications that span technical and social considerations.

Emotion recognition has many issues, and we can confront them with psychological, technical and ethical critiques. From a psychological perspective, Lisa Feldman Barrett et al (2019) challenge assumptions about quantifying and categorizing facial expressions as proxies for emotions on a psychological and social level. They acknowledge some commonalities in, for example, smiling when happy (beyond what might be considered chance). But they also show how the validity of expressions for these and other emotions is severely limited. There is significant variability between contexts and even between people within a given context. There is often significant overlap in which emotions are indicated by a particular expression. And there are many other reasons people act out expressions beyond unconsciously expressing an emotion. The face is a performative stage, not only of a site where emotions are laid bare but of active social and cultural expression.

Emotion recognition can also be challenged from a technical perspective. A review of 'action unit' datasets commonly used to develop emotional expression recognition systems showed major issues in who is included (Pahl et al, 2022). Gender is distributed evenly, but only in a binary classification. Age is weighted within the 22–35 range. Outside this bracket, younger faces tending to be women and older faces men, creating bias at the intersection of age and gender. Meanwhile White faces were disproportionately represented (over 60 per cent) with other groups much less prevalent and often with additional gender disparities, echoing the intersectional concerns of the 'gender shades' study (Buolamwini and Gebru, 2018). Building on contextual and social specificity concerns (Barrett et al, 2019), as well as performative elements of how different people are expected to wear different expressions to match expectations of different social and emotional roles, this is particularly problematic when recognition technologies are ported across contexts as generic technical solutions.

From an ethical perspective we can further challenge emotion recognition. 'Should' they ever be used on people, particularly when they are employing various types of proxy data for emotions (Stark and Hoey, 2021: 785)? This includes questions over the theoretical framings and data types, as well as questions of agency in the design, data and deployment of these systems. More often than not they are used to entrench power rather than challenge it. Luke Stark and Jesse Hoey (2021) raise issues with different types of emotion recognition and the 'troubling norms' (2021: 788) that see external expression, emotion and motivation as being mutually exchangeable proxies. This issue is highly relevant when it comes to the questionable act of judging trustworthiness on faces.

Aside from flaws in technical and psychological assumptions, each facial or emotion recognition technology carries a profound question of whether (and, if at all, when and by whom and on whom) these tools should be used. The chains of quantifiable proxies for people, their emotions, and values such as trust, mean that the wrong questions tend to get asked, and the wrong framings used. The answers then serve only to perpetuate technology discourses that push for expanding these same flawed logics. When emotions and other aspects of identity are perpetually quantified, historical biases and burdens are perpetuated. This includes the gendered burden of bodily surveillance being renewed in workplace datafication (Stark et al, 2020) and the racial histories of surveillance (Browne, 2015) from colonialism through to identifying Black Lives Matter protesters.

In 2022, following the increasing prevalence of these critiques, Microsoft removed access to emotion, age and gender recognition from its facial recognition software. This was a positive step, but with limitations. Justifying the decision making with privacy concerns and a lack of scientific consensus on what emotions are is not the same as making a stand against the power structures of classification and the underlying injustices that can occur. By removing access rather than dismantling the systems, Microsoft still pushes the narrative that it is a matter of technological accuracy rather than a social issue that is at stake; it still seeks to legitimize the larger project of quantifying faces. The areas of application for identification and recognition technologies continue to expand, from borders and policing to transport, public spaces, education, work and shopping. Our entire lives physically, psychologically and socially are shaped according to the decisions of sociotechnical systems entangled with systemic injustice. The goal of quantifying whether a face is trustworthy is tied to trust in technologies, discourses and systems of power.

# TEN

# Conclusion: False Trade-Offs

Throughout this book, I have discussed the role of trust in technology discourses as part of a process of quantification, extraction and legitimization that entrenches existing inequalities and injustices. I have shown how this process, which I have called trustification, consolidates power in the hands of those designing and deploying technologies on populations around the world. Embedded within these narratives is a computational logic that emphasizes the reduction of people and of social problems into something that is countable and can appear solvable. It is through this framing that technological solutionism, and the application of generic solutions across contexts, seeks discursive legitimacy. In the absence of trust, technology discourses measure proxies for trust in order to extract from populations the legitimacy that trust can provide.

Trusting technology is virtually impossible. When we cannot tell who or what it is we are trusting amid sprawling sociotechnical assemblages, trustworthiness becomes replaced by what levels of trust can be measured. In the absence of both specific trust and the possibility of trust, we are expected to act as if we trust anyway. The discourses of technology development and deployment serve to enact what I have called trustification. This is the construction of situations in which populations are forced to perform the conditions of trust when trust itself does not or cannot exist.

Discourses are important in mediating power relations of technologies. As Simone Natale writes, 'technologies function not only at a material and technical level but also through the narrative they generate or into which they are forced' (2021: 55). We must pay close attention to the stories constructed around technologies, how they are constructed, and how those stories emerge from historical narratives such as colonialism, capitalism, heteropatriarchy and others. Throughout this book I have emphasized the performative nature of technologies and their discourses, the way that telling stories constitutes those material and social conditions differently for different populations.

Trustification fuses this discursive process with quantification. We can describe it as a conflating of narrative logics and computational logics. By measuring trust, and by speaking those measurements, trustification constitutes the conditions of trust regardless of whether trustworthiness or trusting relations actually exist. In doing so, it also creates the expectations of trust and the expectation of legitimacy generated by populations performatively acting as if they trust. This operates at the level of specific technical objects, the specific sociotechnical assemblages in which they are embedded, and the broader technology discourses that legitimise the framing and the performative process itself.

In the process of quantification, proxy variables tend to stand in for the nebulous concept of trust and, in doing so, constitute a particular framing of what trust stands for. Proxies are mobilized to embed mainstream discourses with false trade-offs. Innovation vs regulation is the prime example of this. The two are not incompatible, and the very idea of innovation without constraint removes the conditions for needing to be 'innovative'. The two concepts are not opposites, and the way the trade-off is narrated hides the interests of money and power that underpin the priorities of the discourse.

Similarly, the constructed tensions between privacy and security (and both are also traded off against functionality) focus attention on a dichotomy of individual agency and collective concerns, often expressed in opposition to state power or in fear of crime. Yet this trade-off too is a false one, masking the links with corporate and research data practices that exacerbate the unjust targeting of specific marginalized groups: racialized surveillance; the monitoring of women's and trans folk's bodies; or the tracking of queer sexualities. Many of the technologies that support privacy and security are the same, and tensions with functionality can be seen more as issues of design practices and marketing expectations.

False trade-offs de-emphasize or conceal existing power relations and social inequities. Trust is not an inequality of balance between principles. Centring ideas of balance within technology discourses recalls the zero-sum thinking that entrenches an economic or transactional view of human lives. It obfuscates the absence of trustworthiness demonstrated by those in power, limits the potential for transformative justice, and enables the perpetuation of unjust power relations.

Trustification mediates this balance by converting it into a one-directional target that can be gamed from populations. It does this by collapsing complex social relations into a single metric of trust/no trust, risk/benefit, which conceals the fact that it is often the same groups who are harmed at either end of the scale. Those who are over-surveilled (Black and/or trans people, for example) are also often those who have worse access to appropriate healthcare, economic support, agency in regulation, and other tools of addressing social harms.

Trustification is used to elide scrutiny by offering false trade-offs that shift discourse, visibility, attention and effort away from underlying issues of inequality and injustice. Measures of potential harms of technology – values such as fairness, accountability and transparency (Abdurahman, 2019) – are themselves often reduced into proxy variables for trust. When these concepts are reduced to quantifiable targets, we must ask what these measures are for? What is their aim? Is it to extract legitimacy from populations, trustifying shallow terminology as a cover for continued power inequities? Or is it a move against trustification, towards justice in the way AI and other technology systems are designed and deployed? What narratives are constituting the norms and expectations that manage these purposes of technology?

Transparency of data, for example, is often meaningless without transparency of the methods and algorithms through which they are used (Isin and Ruppert, 2020). Features such as transparency should 'not be conceived as widgets that can be imported into a given context to promote trust' (Thornton et al, 2021). Similarly, fairness within data can be limited without fairness in the explanations of how systems work to improve justice in decision making (Balagopalan et al, 2022). This recalls the need for epistemic justice to support data justice.

The push towards data justice is 'fairness in the way people are made visible, represented and treated as a result of their production of data' (Taylor, 2017: 2). We must pay close attention to the discourses that legitimize certain technologies and certain sociotechnical power relations over others. The focus should always remain on questions of who – whose 'good', who trusts, who wants to be trusted, who benefits, who is exploited, who has power, who is excluded, who needs legitimacy, from whom, and who decides how these questions are framed.

There are massive asymmetries between those who are expected to trust and those who acquire or require it, stemming from the problematic conceptions of trust as a measurable value. Those already marginalized by unjust social narratives and power structures are either the ones trust is most extracted from or used as a resource against which to extract trust from privileged groups. This is worsened by the increased burden of effort that falls on those from marginalized groups doing the critical work required to combat injustice, those like Timnit Gebru and Michelle Mitchell who were fired from Google's AI ethics team for asking important questions about the social and environmental impact of the large language models that underpin tools like Google search engine.

The efforts required to constantly critique inequitable systems could rather be spent on generating alternative systems. And yet critical research is often accused of not providing enough solutions. But the work of raising a problem, of framing an issue, can lead to a reframing of the context in which the issue was defined, rather than a singular patch to fix one specific symptom. The aim is systemic change, a disruption to the harmful mainstream discourses of technology. But those discourses always attempt to enforce a logic of solutionism that treats every social problem as a measurable problem solvable by quantification and computation.

The question becomes less about asking whether we can trust a particular technology, and more about asking where trust can and should lie, where responsibility does and should lie, within the sociotechnical assemblage in which a given technology is embedded. Do the loci and directions of trust and responsibility align? Here trust is not something we have but something we do, on the part of the truster and the trustee. This emphasizes trust as a factor of trustworthiness, and with which mistrust is inextricably entwined.

Mistrust is not a negative value, as trustification discourses would suggest. It is a necessary critical frame without which trustworthiness cannot be demonstrated and without which relations of trust are not possible. In the development of AI and other technologies, 'resistance is not a force to fear: it is a powerful signal' (Peppin, 2021: 201). It goes further than revealing areas for gradual improvement or showing what uses of technology are deemed acceptable. Mistrust allows us to shift the question. It helps us ask not only 'can this person, organization or technology be trusted?' but 'how is trust being operationalized to legitimize existing inequalities and power structures?'. Mistrust is a challenge to the conditions of inequality and injustice, a potential for systemic change that reconfigures power relations and the conditions of performing trust rather than simply a counter to it. Mistrust can enable us to perform new norms of technology, to create new contexts in which trust is built socially and collectively rather than extracted numerically and asymmetrically.

Pippa Norris (1999) describes trust as easy to lose but difficult to gain. However, she does not see this as necessarily a bad thing. Trust should have strict requirements on the trustee to prove their trustworthiness. This may involve reforming institutions to demonstrate that they deserve to be trusted (Zuckerman, 2021: 46), baking in more just interests and values in their design and operation. What is more, Norris formulates mistrust as an important part of developing critical citizens. I would add therefore that we do not want to reverse mistrust.

We are not aiming for trust. We need to reconfigure the power relations and discourses behind the trust–mistrust dichotomy. Contextualized mistrust is a better measure of inequality, of untrustworthiness born of systemic injustice, than trust is a measure of faith or confidence in the authority and legitimacy of privilege. This goes along with highlighting the need for qualitative, narrative methods rather than measurement as quantifying trust–mistrust as a sliding scale. By telling different stories of what and who technology is for, we can shift the narrative away from distracting tensions that mediate debates. This helps us look for more transformative ways of reconfiguring sociotechnical systems, of rewriting the discourse.

Mistrust in technology needs repoliticizing, countering the 'offloading of politics' (Andrejevic, 2021) created by computational and economic logics. Discourses that label technology as neutral and objective seek to depoliticize mistrust, to see it as a negative quality in populations rather than an integral part of challenging power. This returns us to the issue of trust in sociotechnical assemblages and looking at why mistrust is there to find points where systemic changes can be made.

There is no simple fix to the issue of trust. It is not a solvable problem but an ongoing series of relations entailing power, responsibility and vulnerability. Trust requires different framings in different contexts and in the different aspects of the sociotechnical assemblages that constitute any given technology and technology discourses in general.

In this book I have shown how trustification works within these discourses and explored the roots of this process across state, corporate, research and media logics of power. Each setting feeds into the overarching discourses of trustification, but each uses different methods and builds on related but different existing narratives, logics and power structures. After tracing the processes and discourses through these different contexts I have applied the concept of trustification to three cross-cutting case studies: COVID-19 tracing apps, focusing on the UK and comparing examples from specific contexts across the Global South; tech for good initiatives, particularly the use of AI and data programs within a humanitarian framing; and the algorithmic assessment of trustworthiness in facial recognition and emotion recognition technologies.

Across these arenas, a recurring theme has been that trust remains an issue. But it is not an issue in terms of a deficit to be filled. Trust is the problem in so far as it continues to be seen as a goal to be achieved. The issue is a lack of mistrust. Against the extractive logics of trustification, mistrust takes on an increasingly important role to enable power for populations, and in particular marginalized groups.

What recommendations or responses can we suggest? Specificity is important. It is necessary to identify within sociotechnical assemblages where trust is placed, where it is needed, who is expected to trust, who is harmed, what is the technology, what ideologies does it perform. Context is also important. In his discussion on the value of mistrust, Ethan Zuckerman states that 'trust is situational' (2021: 22). It is also voluntary and variable, and it involves risk. But the burdens of risk lie unevenly. Trustification seeks to apply generic computational logics to reduce any issue into a quantifiable problem solvable by technology and obfuscate the inequalities and injustices that are made worse by the uneven burdens of technology.

We require an acknowledgement and understanding of specific contexts to create specific measures that can challenge not only the arrangements of power that surround a given technology or setting, but to situate that technology within the particular set of historical narratives that have led to inequalities today. This often includes refusing technological solutions and empowering those most affected to be able to voice their mistrust and make such a refusal. This will need a transformation of the discursive framing within which the norms of ownership, control and agenda setting shape sociotechnical assemblages.

We require a shift in logics and aims. We need not solutions, not extraction, not quantification, but a relational and social approach towards redistribution and transformation. We can look to ideas like 'algorithmic reparations' (Davis et al, 2021) to tackle the root of harms as well as to reconfigure the priorities of how systems can and should be designed going forward.

Metricization needs replacing with politicization. This requires an emphasis on narrative and a protection of the incomputable society. Trust becomes the issue against which mistrust rises to empower those from whom legitimacy is extracted. Mistrust is not the problem, but one answer to the ways technology is used to perpetuate inequalities. By engaging productively with mistrust to challenge power, we can do trust differently, create different norms of technology, and push towards systemic change.

# References

Abbas, R. & Michael, K. (2020) 'COVID-19 contact trace app deployments: learnings from Australia and Singapore', *IEEE Consumer Electronics*, 9(5): 65–70.

Abdurahman, K. (2019) 'FAT* Be Wilin'', *Medium – UpFromTheCracks*, [online] 25 February, https://upfromthecracks.medium.com/fat-be-wilin-deb56bf92539

Abebe, R., Aruleba, K., Birhane, A., Kingsley, S., Obaido, G., Remy, S.L. & Sadagopan, S. (2021) 'Narratives and counternarratives on data sharing in Africa', *FAccT'21*: 329–41.

Ada Lovelace Institute & Traverse (2020) 'Confidence in a crisis? Building public trust in a contact tracing app', *Ada Lovelace Institute*, [online] August, https://www.adalovelaceinstitute.org/wp-content/uploads/2020/08/Ada-Lovelace-Institute_COVID-19_Contact_Tracing_Confidence-in-a-crisis-report-3.pdf

Ahmed, S. (2012) *On Being Included: Racism and Diversity in Institutional Life*, Durham, NC: Duke University Press.

Ahmed, S. (2017) *Living A Feminist Life*, Durham, NC: Duke University Press.

AI HLEG (2019) 'Ethics guidelines for trustworthy AI', European Commission, [online] 8 April, https://ec.europa.eu/futurium/en/ai-alliance-consultation.1.html

Alkhatib, A. (2020) 'Digital contact tracing', [Blog], https://ali-alkhatib.com/blog/digital-contact-tracing

Allen, J. (2022) 'The Integrity Institute's analysis of Facebook's Widely Viewed Content Report [2021-Q4]', *Integrity Institute*, [online] 30 March, https://integrityinstitute.org/widely-viewed-content-analysis-tracking-dashboard

Altmann, S. et al (2020) 'Acceptability of App-based contact tracing for COVID-19: cross-country survey study', *JMIR Mhealth Uhealth*, 8(8): e19857.

Álvarez Ugarte, R. (2020) 'Layers of crises: when pandemics meet institutional and economic havoc', in L. Taylor, G. Sharma, A. Martin & S. Jameson (eds) *Data Justice and Covid-19: Global Perspectives*, Manchester: Meatspace, pp 84–9.

Ammanath, B. (2022) 'Why is solving for trust in AI so challenging?', *Forbes*, [online] 16 May, https://www.forbes.com/sites/forbesbusinesscouncil/2022/05/16/why-is-solving-for-trust-in-ai-so-challenging/.

Amnesty (2020a) 'Qatar: contact tracing app security flaw exposed sensitive personal details of more than one million', *Amnesty International*, [online] 26 May, https://www.amnesty.org/en/latest/news/2020/05/qatar-covid19-contact-tracing-app-security-flaw/

Amnesty (2020b) 'Bahrain, Kuwait and Norway contact tracing apps among most dangerous for privacy', *Amnesty International*, [online] 16 June, https://www.amnesty.org/en/latest/news/2020/06/bahrain-kuwait-norway-contact-tracing-apps-danger-for-privacy/

Amoore, L. (2013) *The Politics of Possibility: Risk and Security Beyond Probability*, Durham, NC: Duke University Press.

Andrejevic, M. (2021) 'Keynote', *Data Justice 2021*, 20–21 May, Cardiff University, UK.

Aouragh, M., Gürses, S., Pritcherd, H. & Snelting, F. (2020) 'The extractive infrastructures of contact tracing apps', *Journal of Environmental Media*, 1(1): 1–9.

Ashwin, J.M. (2014) 'Where in the world is the Internet', [PhD Thesis], University of California Berkeley.

Axios (2022) 'The 2022 Axios Harris Poll 100 reputation rankings', *Axios*, [online] 24 May, https://www.axios.com/2022/05/24/2022-axios-harris-poll-100-rankings

Balagopalan, A., Zhang, H., Hamidieh, K., Hartvigsen, T., Rudzicz, F. & Ghassemi, M.M. (2022) 'The road to explainability is paved with bias: measuring the fairness of explanations', *FAccT'22*: 1–16.

Banerjee, S. (2022) 'Truly trustworthy: a case for trust-optimized AI', *Forbes*, [online] 18 May, https://www.forbes.com/sites/forbestechcouncil/2022/05/18/truly-trustworthy-a-case-for-trust-optimized-ai/

Bareis, J. & Katzenbach, C. (2021) 'Talking AI into being: the narratives and imaginaries of national AI strategies and their performative politics', *Science, Technology & Human Values*: 1–27.

Barrett, L.F., Adolphs, R., Marsella, S., Martinez, A.M. & Pollak, S.D. (2019) 'Emotional expressions reconsidered: challenges to inferring emotion from human facial movements', *Psychological Science in the Public Interest*, 20: 1–68.

Beaugrand, C. (2011) 'Statelessness & administrative violence: Bidūns' survival strategies in Kuwait', *The Muslim World*, 101(2): 228–50.

Becker, M. & Bodó, B. (2021) 'Trust in blockchain-based systems', *Internet Policy Review*, 10(2): 1–10.

Beer, D. (2018) *The Data Gaze: Capitalism, Power and Perception*, London: SAGE.

Benjamin, G. (2020) 'From protecting to performing privacy', *Journal of Sociotechnical Critique*, 1(1): 1–30.

Benjamin, R. (2019) *Race After Technology: Abolitionist Tools for the New Jim Code*, Cambridge: Polity.

Bhambra, G.K. (2014) 'Postcolonial and decolonial dialogues', *Postcolonial Studies*, 17(2): 115–21.

Biesta, G. (2017) 'Education, measurement and the professions: reclaiming a space for democratic professionality in education', *Educational philosophy and theory*, 49(4): 315–30.

Bigo, D. & Bonelli, L. (2019) 'Digital data and the transnational intelligence space', in D. Bigo, E. Isin & E. Ruppert (eds) *Data Politics*, Abingdon: Routledge, pp 100–22.

Birhane, A. (2021a) 'Algorithmic injustice: a relational ethics approach', *Patterns*, 2(2): 1–9.

Birhane, A. (2021b) 'Cheap AI', in F. Kaltheuner (ed) *Fake AI*, Manchester: Meatspace, pp 41–52.

Birhane, A., Prabhu, V.U. & Kahembwe, E. (2021) 'Multimodal datasets: misogyny, pornography, and malignant stereotypes', *arXiv*, 2110.01963: 1–33.

Blum, A. (2012) *Tubes: Behind the Scenes at the Internet*, London: Penguin.

Bodó, B. (2021) 'Mediated trust: a theoretical framework to address the trustworthiness of technological trust mediators', *New Media & Society*, 23(9): 2668–90.

Bodó, B. & Janssen, H. (2021) 'Here be dragons – maintaining trust in the technologized public sector', *Amsterdam Law School Research Paper* 2021(23), *Institute for Information Law Research Paper* (02): 1–22.

Bogost, I. (2015) 'Cathedral of computation', *The Atlantic*, [online] 15 January, https://www.theatlantic.com/technology/archive/2015/01/the-cathedral-of-computation/384300/.

Boswell, C. (2018) *Manufacturing Political Trust: Targets and Performance Management in Public Policy*, Cambridge: Cambridge University Press.

Botero Arcila, B. (2020) 'A human centric framework to evaluate the risks raised by contact-tracing applications', *ICT4Peace, Geneva*, 14 April: 1–17.

Bouk, D. (2015) *How Our Days Became Numbered: Risk and the Rise of the Statistical Individual*, Chicago, IL: University of Chicago Press.

Bowker, G.C. & Star, S.L. (2000) *Sorting Things Out: Classification and Its Consequences*, Cambridge, MA: MIT Press.

Brayne, S. (2020) *Predict and Surveil: Data, Discretion, and the Future of Policing*, Oxford: Oxford University Press.

Browne, S. (2015) *Dark Matters: On the Surveillance of Blackness*, Durham, NC: Duke University Press.

Budge, J. & Bruce, I. (2021) 'Predictions 2022: leaders who embrace trust set the bar for new sustainability and AI goals', *Forrester*, [online] 16 November, https://www.forrester.com/blogs/predictions-2022-leaders-who-embrace-trust-set-the-bar-for-new-sustainability-ai-goals/

Buolamwini, J. & Gebru, T. (2018) 'Gender shades: intersectional accuracy disparities in commercial gender classification', *FAT*★' *18*: 77–91.

Burrell, J. (2016) 'How the machine "thinks": understanding opacity in machine learning algorithms', *Big Data & Society*, 3(1): 1–12.

Butler, J. (1990) *Gender Trouble: Feminism and the Subversion of Identity*, Abingdon: Routledge.

Butler, J. (2018) *Notes Toward a Performative Theory of Assembly*, Cambridge, MA: Harvard University Press.

Cameron, W. (1963) *Informal Sociology: A Casual Introduction to Sociological Thinking*, New York, NY: Random House.

Carmi, E. (2020) *Media Distortions: Understanding the Power Behind Spam, Noise, and Other Deviant Media*, New York, NY: Peter Lang.

Carmi, E. (2021) 'A feminist critique to digital consent', *Seminar.net*, 17(2), https://doi.org/10.7577/seminar.4291

Chatterjee, P. & Maira, S. (2014) *The Imperial University: Academic Repression and Scholarly Dissent*, Minneapolis, MN: University of Minnesota Press.

Chiapello, E. (2007) 'Accounting and the birth of the notion of capitalism', *Critical Perspectives on Accounting*, 18: 263–96.

Cini, M. & Czulno, P. (2022) 'Digital Single Market and the EU Competition Regime: an explanation of policy change', *Journal of European Integration*, 44(1): 41–57.

Coady, D. (2017) 'Epistemic injustice as distributive injustice', in I.J. Kidd, J. Medina & G. Pohlhaus (eds) *Routledge Handbook of Epistemic Injustice*, Abingdon: Routledge, pp 61–8.

Cormack, D. & Kukutai, T. (2021) 'Pandemic paternalism: a reflection on indigenous data from Aotearoa', in S. Milan, E. Treré & S. Masiero (eds) *COVID-19 from the Margins*, Amsterdam: Institute of Network Cultures, pp 141–4.

Costanza-Chock, S. (2020) *Design Justice: Community-Led Practices to Build the Worlds We Need*, Cambridge, MA: MIT Press.

Couldry, N. & Mejias, U.A. (2019) 'Data colonialism: rethinking big data's relation to the contemporary subject', *Television & New Media*, 20(4): 336–49.

Couldry, N. & Mejias, U.A. (2021) 'The decolonial turn in data and technology research: what is at stake and where is it heading?', *Information, Communication & Society*: 1–17.

Cruz-Santiago, A. (2020) 'Normalising digital surveillance', in L. Taylor, G. Sharma, A. Martin & S. Jameson (eds) *Data Justice and Covid-19: Global Perspectives*, Manchester: Meatspace, pp 184–9.

Das, S. (2021) 'Surveillance in the time of Covid-19: the case of the Indian contact tracing app Aarogya Setu', in S. Milan, E. Treré & S. Masiero (eds) *COVID-19 from the Margins*, Amsterdam: Institute of Network Cultures, pp 57–60.

Davis, J.L., Williams, A. & Yang, M.W. (2021) 'Algorithmic reparation', *Big Data & Society*, 8(2): 1–12.

Delacroix, S. & Lawrence, N. (2019) 'Bottom-up data trusts: disturbing the "one size fits all" approach to data governance', *International Data Privacy Law*, 9(4): 236–52.

Delgado, D. (2021) 'The dark database: facial recognition and its "failure" to enroll', *Media-N*, 17(2): 69–79.

Devine, D., Gaskell, J., Jennings, W. & Stoker, G. (2020) 'Trust and the coronavirus pandemic: what are the consequences of and for trust? An early review of the literature', *Political Studies Review*, 19(2): 274–85, https://doi.org/10.1177/1478929920948684

D'Ignazio, C. and Klein, L. (2020a) *Data Feminism*, Cambridge, MA: MIT Press.

D'Ignazio, C. & Klein, L. (2020b) 'Seven intersectional feminist principles for equitable and actionable COVID-19 data', *Big Data & Society*: 1–6.

Dourish, P. & Gómez Cruz, E. (2018) 'Datafication and data fiction: narrating data and narrating with data', *Big Data & Society*: 1–10.

Downey, A. (2020) 'NHS Covid-19 contact-tracing app launched in England and Wales', *Digital Health*, [online] 24 September, https://www.digitalhealth.net/2020/09/nhs-covid-19-contact-tracing-app-launched-england-wales/

Eder, C., Mochmann, I.C. & Quandt, M. (2014) 'Editor's introduction: political trust and political disenchantment in a comparative perspective', in C. Eder, I.C. Mochmann & M. Quandt (eds) *Political Trust and Disenchantment with Politics: International Perspectives*, Leiden: Brill, pp. 1–18.

Elish M.C. & boyd, d. (2018) 'Situating methods in the magic of big data and AI', *Communication Monographs*, 85(1): 57–80.

Ellcessor, E. (2022) *In Case of Emergency: How Technologies Mediate Crisis and Normalize Inequality*, New York, NY: NYU Press.

Engelmann, S., Ullstein, C., Papakyriakopoulos, O. & Grossklags, J. (2022) 'What people think AI should infer from faces', *FAccT'22*: 1–12.

European Commission (2020) 'White paper on artificial intelligence – a European approach to excellence and trust (White Paper COM(2020) 65 final)', *European Commission*, [online] 19 February, https://eur-lex.europa.eu/legal-content/en/ALL/?uri=CELEX:52020DC0065

Fourcade, M. (2011) 'Cents and sensibility: economic valuation and the nature of "nature"', *American Journal of Sociology*, 116(6): 1721–77.

Fricker, M. (2007) *Epistemic Injustice: Power and the Ethics of Knowing*, Oxford: Oxford University Press.

FT (2021) 'Why building trust in AI is essential', *Financial Times*, [online] 2 March, https://www.ft.com/content/85b0882e-3e93-42e7-8411-54f4e24c7f87

Gandy, O.H. (2010) 'Engaging rational discrimination: exploring reasons for placing regulatory constraints on decision support systems', *Ethics and Information Technology*, 12(1): 29–42.

Gandy, O.H. (2021 [1993]) *The Panoptic Sort: A Political Economy of Personal Information* (2nd edn), Oxford: Oxford University Press.

Ganesh, M.I. & Moss, E. (2022) 'Resistance and refusal to algorithmic harms: Varieties of "knowledge projects"', *Media International Australia*: 1–17.

Gekker, A. & Ben-David, A. (2021) 'Data cudgel or how to generate corona-compliance in Israel', in S. Milan, E. Treré, & S. Masiero (eds) *COVID-19 from the Margins*, Amsterdam: Institute of Network Cultures, pp 149–52.

Gillespie, T. (2020) 'Content moderation, AI, and the question of scale', *Big Data & Society,* July–December: 1–5.

Gilliard, C. & Culik, C. (2016) 'Digital redlining, access and privacy', *Common Sense Education*, [online] 24 May, https://www.commonsense.org/education/articles/digital-redlining-access-and-privacy

Gilliard, C. & Golumbia, D. (2021) 'Surveillance as luxury', *RealLife*, [online] 6 July, https://reallifemag.com/luxury-surveillance/

Gilmore, J.N. (2021) 'Predicting Covid-19: wearable technology and the politics of solutionism', *Cultural Studies*, 35(2–3): 382–91.

Glaeser, E.L, Laibson, D.I., Scheinkman, J.A. & Soutter, C.L. (2000) 'Measuring trust', *The Quarterly Journal of Economics*, 115(3): 811–46.

Gonzaga, E. (2021) 'Zombie capitalism and coronavirus time', *Cultural Studies*, 35(2–3): 444–51.

Gorwa, R., Binns, R. & Katzenbach, C. (2020) 'Algorithmic content moderation: technical and political challenges in the automation of platform governance', *Big Data & Society*, January–June: 1–15.

Goubin, S. & Kumlin, S. (2022) 'Political trust and policy demand in changing welfare states: building normative support and easing reform acceptance?', *European Sociological Review*, 38(4): 590–604 https://doi.org/10.1093/esr/jcab061

GPAI (2022) 'Enabling data sharing for social benefit through data trusts', *GPAI*, [online] February, https://gpai.ai/projects/data-gov ernance/data-trusts/

Greenleaf, G. & Kemp, K. (2020) 'Australia's 'COVIDSafe App': an experiment in surveillance, trust and law', *University of New South Wales Law Research Series*, 999: 1–17.

Gupte, J. & Mitlin, D. (2021) 'COVID-19: what is not being addressed', *Environment and Urbanization*, 33(1): 211–28.

Guardian (2022) 'The uber files', *The Guardian*, [online] 11 July, https://www.theguardian.com/news/series/uber-files

Habitat3 (2016) 'The new urban agenda', *Habitat3*, [online] 20 October, https://habitat3.org/the-new-urban-agenda/

Hacking, I. (1990) *The Taming of Chance*, Cambridge: Cambridge University Press.

Hall, G. (2016) *The Uberfication of the University*, Minneapolis, MN: University of Minnesota Press.

Hamidi, F., Scheuerman, M.K. & Branham, S.M. (2018) 'Gender recognition or gender reductionism? The social implications of embedded gender recognition systems', *CHI' 18*: 1–13.

Hamraie, A. & Fritsch, K. (2019) 'Crip technoscience manifesto', *Catalyst: Feminism, Theory, Technoscience*, 5(1): 1–33.

Hanna, A., Denton, E., Smart, A. & Smith-Loud, J. (2020) 'Towards a critical race methodology in algorithmic fairness', *FAccT'20*: 501–12.

Haraway, D. (1988) 'Situated knowledges: the science question in feminism and the privilege of partial perspective', *Feminist Studies*, 14(3): 575–99.

Hardin, R. (2006) *Trust*, Cambridge: Polity.

Hartley, K. & Jarvis, D.S.L. (2020) 'Policymaking in a low-trust state: legitimacy, state capacity, and responses to COVID-19 in Hong Kong', *Policy and Society*, 39(3): 403–23.

Hartman, S. (2006) *Losing Your Mother: A Journey Along the Atlantic Slave Route*, London: Serpent's Tail.

Hawlitschek, F., Notheisen, B. & Teubner, T. (2018) 'The limits of trust-free systems: a literature review on blockchain technology and trust in the sharing economy', *Electronic Commerce Research and Applications,* 29: 50–63.

Hawlitschek, F., Teubner, T. & Weinhardt, C. (2016) 'Trust in the sharing economy', *Die Unternehmung*, 70(1): 26–44.

Hicks, M. (2021) 'When did the fire start?', in T.S. Mullaney, B. Peters, M. Hicks & K. Philip (eds) *Your Computer is on Fire*, Cambridge, MA: MIT Press, pp. 11–28.

Hinch, R. et al (2020) 'Effective configurations of a digital contact tracing app: a report to NHSX', *Oxford University*, [online] 16 April, https://045.medsci.ox.ac.uk/files/files/report-effective-app-configurations.pdf

Hoffmann, A.L. (2018) 'Data violence and how bad engineering choices can damage society', *Medium*, [online] 30 April, https://medium.com/s/story/data-violence-and-how-bad-engineering-choices-can-damage-society-39e44150e1d4

Hoffmann, A.L. (2021) 'Terms of inclusion: data, discourse, violence', *New Media & Society*, 23(12): 3539–56.

Hoffmann, A.L., Proferes, N. & Zimmer, M. (2018) '"Making the world more open and connected": Mark Zuckerberg and the discursive construction of Facebook and its users', *New Media & Society*, 20(1): 199–218.

Holmwood, J. (2016) 'Papering over the cracks: the coming white paper and the dismantling of higher education', *Campaign for the Public University*, [online] 25 April, http://publicuniversity.org.uk/2016/04/25/papering-over-the-cracks-the-green-paper-and-the-stratification-of-higher-education

Hooghe, M., Marien, S. & Oser, J. (2017) 'Great expectations: the effect of democratic ideals on political trust in European democracies', *Contemporary Politics*, 23(2): 214–30.

Hookway, C. (2010) 'Some varieties of epistemic injustice: reflections on Fricker', *Episteme*, 7(2): 151–63.

Icaza, R. & Vasquez, R. (2018) 'Diversity or decolonisation? Researching diversity at the University of Amsterdam', in G.K. Bhambra, D. Gebrial & K. Nisancioglu (eds) *Decolonising the University*, London: Pluto Press, pp 108–28.

Irani, L., Vertesi, J., Dourish, P., Philip, K. & Grinter, R.E. (2010) 'Postcolonial computing: a lens on design and development', *CHI' 10*: 1311–20.

Isin, E. and Ruppert, E. (2019) 'Data's empire: postcolonial data politics', in D. Bigo, E. Isin & E. Ruppert (eds) *Data Politics*, Abingdon: Routledge, pp. 207–28.

Isin, E., and Ruppert, E. (2020) 'The birth of sensory power: how a pandemic made it visible?', *Big Data & Society*, July–December: 1–15.

Jacobs, M. (2021) 'How implicit assumptions on the nature of trust shape the understanding of the blockchain technology', *Philosophy & Technology*, 34: 573–87.

Jansen, F. & Cath, C. (2021) 'Algorithmic registers and their limitation as a governance practice', in F. Kaltheuner (ed) *Fake AI*, Manchester: Meatspace, pp 183–92.

Johns, F. (2023) *#Help: Digital Humanitarianism and the Remaking of International Order*, Oxford: Oxford University Press.

Johnson, K. (2021) 'Google targets AI ethics lead Margaret Mitchell after firing Timnit Gebru', *VentureBeat*, [online] 20 January, https://venturebeat.com/2021/01/20/google-targets-ai-ethics-lead-margaret-mitchell-after-firing-timnit-gebru/

Kaltheuner, F. (2021) *Fake AI*, Manchester: Meatspace.

Katzenbach, C. (2021) '"AI will fix this" – the technical, discursive, and political turn to AI in governing communication', *Big Data & Society*, 8(2): 1–8.

Kaurin, D. (2020) 'The dangers of digital contact tracing: lessons from the HIV pandemic', in L. Taylor, G. Sharma, A. Martin & S. Jameson (eds) *Data Justice and Covid-19: Global Perspectives*, Manchester: Meatspace, pp 64–9.

Kazansky, B. (2015) 'FCJ-195 privacy, responsibility, and human rights activism', *The Fibreculture Journal*, 26: 190–208.

Kennedy, H., Steedman, R. & Jones, R. (2022) 'Researching public trust in datafication: reflections on the deliberative citizen jury as method', in A. Hepp, J. Jarke & L. Kramp (eds) *New Perspectives in Critical Data Studies: The Ambivalences of Data Power*, Cham: Palgrave Macmillan, pp 391–414.

Keyes, O. (2018) 'The misgendering machines: trans/HCI implications of automatic gender recognition', *CSCW' 18*: 1–22.

Keyes, O. (2020) 'Who counts? Contact tracing and the perils of privacy', in L. Taylor, G. Sharma, A. Martin, & S. Jameson (eds) *Data Justice and Covid-19: Global Perspectives*, Manchester: Meatspace, pp 58–63.

Kim, Y. (2021) 'Bio or Zoe?: dilemmas of biopolitics and data governmentality during COVID-19', *Cultural Studies*, 35(2–3): 370–81.

Kim, Y., Chen, Y. & Liang, F. (2021) 'Engineering care in pandemic technogovernance: the politics of care in China and South Korea's COVID-19 tracking apps', *New Media & Society*: 1–19.

King, L.G.J. (2014) 'More than slaves: Black Founders, Benjamin Banneker, and critical intellectual agency', *Social Studies Research and Practice*, 9(3): 88–105.

König, P.D. (2017) 'The place of conditionality and individual responsibility in a "data-driven economy"', *Big Data & Society*: 1–14.

Lapuste, M. (2015) 'Cloud-computing services. Legal dimension of use and development of this type of services', *National Strategies Observer*, 2(2): 124–7.

Lawrence, N. (2016) 'Data trusts could allay our privacy fears', *The Guardian*, [online] 3 June, https://www.theguardian.com/media-network/2016/jun/03/data-trusts-privacy-fears-feudalism-democracy

Levina, M. & Hasinoff, A.A. (2017) 'The Silicon Valley ethos: tech industry products, discourses, and practices', *Television & New Media*, 18(6): 489–95.

Limaye, R.J., Sauer, M., Ali, J., Bernstein, J., Wahl, B. & Barnhill, A. (2020) 'Building trust while influencing online COVID-19 content in the social media world', *The Lancet: Digital Health*, 2(6): 277–8.

Liu, J.H., Milojev, P., de Zúñiga, H.G. & Zhang, R.J. (2018) 'The Global Trust Inventory as a "proxy measure" for social capital: measurement and impact in 11 democratic societies', *Journal of Cross-Cultural Psychology*, 49(5): 789–810.

Livio, M. & Emerson, L. (2019) 'Towards feminist labs: provocations for collective knowledge-making', in L. Bogers & L. Chiappini (eds) *The Critical Makers Reader: (Un)learning Technology*, Amsterdam: Institute of Network Cultures, pp 286–97.

Llansó, E.J. (2020) 'No amount of "AI" in content moderation will solve filtering's prior restraint problem', *Big Data & Society*, January–June: 1–6.

Lucero, V. (2020) 'Fast tech to silence dissent, slow tech for public health crisis ', in L. Taylor, G. Sharma, A. Martin & S. Jameson (eds) *Data Justice and Covid-19: Global Perspectives*, Manchester: Meatspace, pp 224–31.

Lyon, D. (2019) 'Surveillance capitalism, surveillance culture and data politics', in D. Bigo, E. Isin, & E. Ruppert (eds) *Data Politics: Worlds, Subjects, Rights*, Abingdon: Routledge, pp 64–78.

Mackenzie, A. (2015) 'The production of prediction: what does machine learning want?', *European Journal of Cultural Studies*, 18(4–5): 429–45.

Madianou, M. (2021) 'Nonhuman humanitarianism: when "AI for good" can be harmful', *Information, Communication & Society*, 24(6): 850–68.

Madianou, M., Ong, J.C., Longboan, L. & Cornelio, J.S. (2016) 'The appearance of accountability: communication technologies and power asymmetries in humanitarian aid and disaster recovery', *Journal of Communication*, 66(6): 960–81.

Maris, E. & Baym, N. (2022) 'Community rankings and affective discipline: the case of fandometrics', in A. Hepp, J. Jarke & L. Kramp (eds) *New Perspectives in Critical Data Studies: The Ambivalences of Data Power*, Cham: Palgrave Macmillan, pp. 323–44.

Marlinspike, M. (2022) 'My first impressions of Web3', *Moxie*, [online] 7 January, https://moxie.org/2022/01/07/web3-first-impressions.html

Martin, A. (2023) 'Aidwashing surveillance: critiquing the corporate exploitation of humanitarian crises', *Surveillance & Society*, 21(1): 96–102.

Masiero, S., Milan, S. & Treré, E. (2021) 'COVID-19 from the margins: crafting a (cosmopolitan) theory', *Global Media Journal – German Edition*, 11(1): 1–23.

Maxigas, P. & Latzko-Toth, G. (2020) 'Trusted commons: why 'old' social media matter', *Internet Policy Review*, 9(4): 1–20.

Mayer- Schönberger, V. and Cukier, K. (2013) *Big Data*, London: John Murray.

Mbembe, A. (2019) *Necropolitics*, Durham, NC: Duke University Press.

Medina, J. (2007) *Speaking from Elsewhere: A New Contextualist Perspective on Meaning, Identity, and Discursive Agency*, New York, NY: SUNY Press.

Medina, J. (2012) *The Epistemology of Resistance: Gender and Racial Oppression, Epistemic Injustice, and Resistant Imaginations*, Oxford: Oxford University Press.

Mejias, U.A. & Couldry, N. (2019) 'Datafication', *Internet Policy Review*, 8(4): 1–10.

Metcalfe, P. & Dencik, L. (2019) 'The politics of big borders: data (in)justice and the governance of refugees', *First Monday*, 24(4), https://doi.org/10.5210/fm.v24i4.9934

Mezei, P. & Verteş-Olteanu, A. (2020) 'From trust in the system to trust in the content', *Internet Policy Review*, 9(4): 1–28.

Mhlambi, S. (2020) 'From rationality to relationality: Ubuntu as an ethical and human rights framework for artificial intelligence governance', *Carr Center Discussion Paper Series*, 2020-009, https://carrcenter.hks.harvard.edu/files/cchr/files/ccdp_2020-009_sabelo_b.pdf

Milan, S. (2020) 'Techno-solutionism and the standard human in the making of the COVID-19 pandemic', *Big Data & Society*: 1–7.

Milan, S. (2022) 'Counting, debunking, making, witnessing, shielding: what critical data studies can learn from data activism during the pandemic', in A. Hepp, J. Jarke & L. Kramp (eds) *New Perspectives in Critical Data Studies: The Ambivalences of Data Power*, Cham: Palgrave Macmillan, pp 445–67.

Milan, S. & Treré, E. (2019) 'Big data from the south(s): beyond data universalism', *Television & New Media*, 20(4), 319–35.

Milan, S., Treré, E. & Masiero, S. (eds) (2021) *COVID-19 from the Margins: Pandemic Invisibilities, Policies and Resistance in the Datafied Society*, Amsterdam: Institute of Network Cultures.

Milne, G. (2021) 'Uses (and abuses) of hype', in F. Kaltheuner (ed) *Fake AI*, Manchester: Meatspace.

Mohamed, S., Png, M.-T. & Isaac, W. (2020) 'Decolonial AI: decolonial theory as sociotechnical foresight in artificial intelligence', *Philosophy and Technology*, 33(4): 659–84.

Morozov, E. (2014) *To Save Everything, Click Here: The Folly of Technological Solutionism*, New York, NY: PublicAffairs.

Mosco, V. (2004) *The Digital Sublime: Myth, Power, and Cyberspace*, Cambridge, MA: The MIT Press.

Mulligan, D. & Bamberger, K. 'Saving governance-by-design', *California Law Review*, 106(3): 697–784.

Mumford, D. (2021) 'Data colonialism: compelling and useful, but whither epistemes?', *Information, Communication & Society*: 1–6.

Murphy, M. (2016) *The Economization of Life*, Durham, NC: Duke University Press.

Natale, S. (2021) *Deceitful Media: Artificial Intelligence and Social Life After the Turing Test*, Oxford: Oxford University Press.

Neilson, B. & Rossiter, N. (2019) 'Thesis on automation and labour', in D. Bigo, E. Isin & E. Ruppert (eds) *Data Politics: Worlds, Subjects, Rights*, Abingdon: Routledge, pp 187–206.

Newton, K. & Norris, P. (2000) 'Confidence in public institutions: faith, culture, or performance?', in S. Pharr, & R. Putnam (eds) *Disaffected Democracies*, Princeton, NJ: Princeton University Press, pp 52–73.

Noble, S. (2018) *Algorithms of Oppression*, New York, NY: NYU Press.

Nooteboom, B. (2002) *Trust: Forms, Foundations, Functions, Failures and Figures*, Cheltenham: Edward Elgar.

Nooteboom, B. (2013) 'Trust and innovation', in R. Bachmann & A. Zaheer (eds) *Handbook of Advances in Trust Research*, Cheltenham: Edward Elgar, pp 106–23.

Norris, P. (1999) *Critical Citizens: Global Support for Democratic Government*, Oxford: Oxford University Press.

Nouwens, M., Liccadri, I., Veale, M., Karger, D. & Kagal, L. (2020) 'Dark patterns after the GDPR: scraping consent pop-ups and demonstrating their influence', *CHI' 20*: 1–13.

Nunn, N. & Wantchekon, L. (2011) 'The slave trade and the origins of mistrust in Africa', *American Economic Review*, 101(7): 3221–52.

OCHA (2022) 'The state of open humanitarian data 2022: assessing data availability across humanitarian crises', *UN Office for the Coordination of Humanitarian Affairs*, [online] January, https://centre.humdata.org/stateofdata2022

O'Donnell, C. (2014) 'Getting played: gamification, bullshit, and the rise of algorithmic surveillance', *Surveillance & Society*, 12(3): 349–59.

OECD (2017) *OECD Guidelines on Measuring Trust*, Paris: OECD.

Onokwue, A. (2020) 'The BackEnd: meet the "Palantir of Africa"', *Tech Cabal*, [online] 19 November, https://techcabal.com/2020/11/19/the-backend-analytics-intelligence-palantir-africa/

Oxford University (2022) 'Making fairer decisions: the bias detection tool developed at Oxford and implemented by Amazon', *Oxford University Social Sciences*, [online], https://www.socsci.ox.ac.uk/making-fairer-decisions-the-bias-detection-tool-developed-at-oxford-and-implemented-by-amazon

Özden-Schilling, C. (2016) 'The infrastructure of markets: from electric power to electronic data', *Economic Anthropology*, 3(1): 68–80.

Pahl, J., Rieger, I., Möller, A., Wittenberg, T. & Schmid, U. (2022) 'Female, white, 27? Bias evaluation on data and algorithms for affect recognition in face', *FAccT'22*: 1–15.

Patel, G. (2006) 'Risky subjects: insurance, sexuality and capital', *Social Text*, 24(4): 25–65.

Peppin, A. (2021) 'The power of resistance: from plutonium rods to silicon chips', in F. Kaltheuner (ed) *Fake AI*, Manchester: Meatspace, pp 193–201.

Poetranto, I. & Lau, J. (2021) 'Covid-19 and its impact on marginalised communities in Singapore, South Korea, Indonesia and the Philippines', in S. Milan, E. Treré & S. Masiero (eds) *COVID-19 from the Margins*, Amsterdam: Institute of Network Cultures, pp 95–101.

Poon, M. (2009) 'From new deal institutions to capital markets: commercial consumer risk scores and the making of subprime mortgage finance', *Accounting, Organisations and Society*, 34(5): 654–74.

Popiel, P. (2018) 'The tech lobby: tracing the contours of new media elite lobbying power, communication', *Culture and Critique*, 11(4): 566–85.

Porter, T. (1995) *Trust in Numbers: The Pursuit of Objectivity in Science and Public Life*, Princeton, NJ: Princeton University Press.

Postman, N. (1993) *Technopoly: The Surrender of Culture to Technology*, New York, NY: Vintage Books.

Quijano, A. (2007) 'Coloniality and modernity/rationality', *Cultural Studies*, 21(2): 168–78.

Raghunath, P. (2021) 'Covid-19 and non-personal data in the Indian context: on the normative ideal of public interest', in S. Milan, E. Treré & S. Masiero (eds) *COVID-19 from the Margins*, Amsterdam: Institute of Network Cultures, pp 200–2.

Raji, I.D. (2021) 'The bodies underneath the rubble', in F. Kaltheuner (ed) *Fake AI*, Manchester: Meatspace, pp 53–62.

Raji, I.D. & Buolamwini, J. (2019) 'Actionable auditing: investigating the impact of publicly naming biased performance results of commercial AI products', *AIES'19*: 429–35.

Raji, I.D., Kumar, I.E., Horowitz, A. & Selbst, A.D. (2022) 'The fallacy of AI functionality', *FAccT'22*: 959–72.

Ralph, M. (2018) 'The value of life: insurance, slavery, and expertise', in S. Beckert & C. Desan (eds) *American Capitalism: New Histories*, New York, NY: Columbia University Press, pp 257–82.

Reia, J. & Cruz, L.F. (2021) 'Seeing through the smart city narrative: data governance, power relations, and regulatory challenges in Brazil', in B. Haggart, N. Tusikov & J.A. Scholte (eds) *Power and Authority in Internet Governance: Return of the State?*, Abingdon: Routledge, pp 219–42.

Ricaurte, P. (2019) 'Data epistemologies, the coloniality of power, and resistance', *Television & New Media*, 20(4): 350–65.

Riekeles, G. (2022) 'I saw first-hand how US tech giants seduced the EU – and undermined democracy', *The Guardian*, [online] 28 June, https://www.theguardian.com/commentisfree/2022/jun/28/i-saw-first-hand-tech-giants-seduced-eu-google-meta

Roberge, J., Senneville, M. & Morin, K. (2020) 'How to translate artificial intelligence? Myths and justifications in public discourse', *Big Data & Society*, 7(1): 1–13.

Rodrigues, L.L., Craig, R.J., Schmidt, P. & Santos, L.J. (2015) 'Documenting, monetising and taxing Brazilian slaves in the eighteenth and nineteenth centuries', *Accounting History Review*, 25(1): 34–67.

Rodríguez, C. (2017) 'Studying media at the margins: Learning from the field', in V. Pickard & G. Yang (eds) *Media Activism in the Digital Age*, Abingdon: Routledge, pp. 49–60.

Ross, C. & Jang, S.J. (2000) 'Neighborhood disorder, fear, and mistrust: the buffering role of social ties with neighbors', *American Journal of Community Psychology,* 28: 401–20.

Ross, C., Mirowsky, J. & Pribesh, S. (2001) 'Powerlessness and the amplification of threat: neighborhood disadvantage, disorder, and mistrust', *American Sociological Review*, 66(4): 568–91.

Roth, W.D. (2016) 'The multiple dimensions of race', *Ethnic and Racial Studies*, 39(8): 1310–38.

Ryan, M. (2020) 'In AI we trust: ethics, artificial intelligence, and reliability', *Science and Engineering Ethics*, 26: 2749–67.

Sadowski, J. (2019) 'When data is capital: datafication, accumulation, and extraction', *Big Data & Society*, 6(1): 1–12.

Sayer, D. (2015) *Rank Hypocrisies: The Insult of the REF*, New York, NY: SAGE.

Scheiber, N. & Conger, K. (2020) 'The great Google revolt', *New York Times*, [online] 18 February, https://www.nytimes.com/interactive/2020/02/18/magazine/google-revolt.html

Scheman, N. (2011) *Shifting Ground: Knowledge and Reality, Transgression and Trustworthiness*, Oxford: Oxford University Press.

Scheuerman, M.K., Pape, M. & Hanna, A. (2021) 'Auto-essentialization: gender in automated facial analysis as extended colonial project', *Big Data & Society*, July–December: 1–15.

Schrape, N. (2014) 'Gamification and governmentality', in M. Fuchs, S. Fizek, P. Ruffino & N. Schrape (eds) *Rethinking Gamification*, Lüneburg: Meson Press, pp 21–45.

Scott, J. (1998) *Seeing Like a State: How Certain Schemes to Improve the Human Condition Have Failed*, Yale, CN: Yale University Press.

Shannon, D. et al (2021) 'Connected and autonomous vehicle injury loss events: potential risk and actuarial considerations for primary insurers', *Risk Management and Insurance Review*, 24(1): 5–35.

Sinders, C. & Hendrix, J. (2021) 'Taking action on dark patterns', *Tech Policy Press*, [online] 26 April, https://techpolicy.press/taking-action-on-dark-patterns/

Singler, B. (2020) 'The AI creation meme: a case study of the new visibility of religion in artificial intelligence discourse', *Religions*, 11(5): 253.

Song, C. & Lee, J. (2016) 'Citizens' use of social media in government, perceived transparency, and trust in government', *Public Performance & Management Review,* 39(2): 430–53.

Spade, D. (2015) *Normal Life: Administrative Violence, Critical Trans Politics and the Limits of Law*, Durham, NC: Duke University Press.

Spanton, R. & Guest, O. (2022) 'Measuring trustworthiness or automating physiognomy? A comment on Safra, Chevallier, Grèzes, and Baumard (2020)', *arXiv*, 2202.08674: 1–3.

Spivak, G.C. (1988) 'Can the subaltern speak?', in C. Nelson & L. Grossberg (eds) *Marxism and the Interpretation of Culture*, London: Macmillan, pp 24–8.

Stanton, B. & Jensen, T. (2021) 'Trust and artificial intelligence, NIST interagency/internal report (NISTIR)', *National Institute of Standards and Technology*, [online] 2 March, https://tsapps.nist. gov/publication/get_pdf.cfm?pub_id=931087.

Stark, L. & Hoey, J. (2021) 'The ethics of emotion in artificial intelligence systems', *FAccT'21*: 782–93.

Stark, L. & Hutson, J. (2022) 'Physiognomic artificial intelligence', *Fordham Intellectual Property, Media and Entertainment Law Journal*, 32(4): 922–78.

Stark, L., Stanhaus, A. & Anthony, D.L. (2020) '"I don't want someone to watch me while I'm working": gendered views of facial recognition technology in workplace surveillance', *Journal of the Association of Information Science Technology*, 71(9): 1074–88.

Striphas, T. (2015) 'Algorithmic culture', *European Journal of Cultural Studies*, 18(4–5): 395–412.

Taylor, L. (2017) 'What is data justice? The case for connecting digital rights and freedoms globally', *Big Data & Society*, 4(2): 1–14.

Teubner, T., Hawlitschek, F. & Adam, M. (2019) 'Reputation transfer', *Business Information Systems Engineering*, 61(2): 229–35.

Thatcher, J., O'Sullivan, D. & Mahmoudi, D. (2016) 'Data colonialism through accumulation by dispossession: new metaphors for daily data', *Environment and Planning D: Society and Space*, 34(6): 990–1006.

Thompson, K. (1984) 'Reflections on trusting trust', *Communications of the ACM*, 27(8): 1–3.

Thornton, L., Knowles, B. & Blair, G. (2021) 'Fifty shades of grey: in praise of a nuanced approach towards trustworthy design', *FAccT'21*: 64–76.

Tréguer, F. (2019) 'Seeing like Big Tech: security assemblages, technology, and the future of state bureaucracy', in D. Bigo, E. Isin & E. Ruppert (eds) Data Politics: *Worlds, Subjects, Rights*, Abingdon: Routledge, pp 145–64.

Tufekci, Z. (2014) 'Engineering the public: Big Data, surveillance and computational politics', *First Monday*, 19(7).

Tuhiwai Smith, L. (2012) *Decolonizing Methodologies: Research and Indigenous Peoples*, London: Zed Books.

Tyler, T.R. (1998) 'Public mistrust of the law: political perspective', *University of Cincinnati Law Review*, 66(3): 847–76.

Vaidhyanathan, S. (2011) *The Googlization of Everything (And Why We Should Worry)*, Berkeley, CA: University of California Press.

Van Bavel, J.J., Baicker, K, Boggio, P.S. et al (2020) 'Using social and behavioural science to support COVID-19 pandemic response', *Nature Human Behaviour*, 4: 460–71.

Van Dijk, J. (2014) 'Datafication, dataism and dataveillance: Big Data between scientific paradigm and ideology', *Surveillance & Society*, 12(2): 197–208.

Viljoen, S. (2021) 'Data relations', *Logic*, 13, [online] 17 May, https://logicmag.io/distribution/data-relations/

Volti, R. (2014) *Society and Technological Change*, Duffield: Worth Publishers.

UKCDR (2022) 'Covid-19 research project tracker by UKCDR & GloPID-R', *UKCDR,* [online] 22 April, https://ukcdr.org.uk/covid-circle/covid-19-research-project-tracker/

W3C (2012) 'Mission', *W3C*, [online] 29 August, https://w3.org/Consortium/mission

Weheliye, A.G. (2014) *Habeas Viscus: Racializing Assemblages, Biopolitics, and Black Feminist Theories of the Human*, Durham, NC: Duke University Press.

Werbach, K. (2018) *The Blockchain and the New Architecture of Trust*, Cambridge, MA: MIT Press.

Whitson, J. (2015) 'Foucault's fitbit: governance and gamification', in S.P. Walz & S. Deterding (eds) *Gameful Worlds: Approaches, Issues, Applications*, Cambridge, MA: MIT Press, pp 339–58.

Whitson, J. & Simon, B. (2014) 'Game studies meets surveillance studies at the edge of digital culture', *Surveillance & Society*, 12(3): 309–19.

Wolfe, R. & Caliskan, A. (2022) 'Markedness in visual semantic AI', *FAccT'22*: 1–11.

Wolfe, R., Banaji, M.R. & Caliskan, A. (2022) 'Evidence for hypodescent in visual semantic AI', *FAccT'22*: 1–12.

Woodall, A. & Ringel, S. (2020) 'Blockchain archival discourse: trust and the imaginaries of digital preservation', *New Media & Society*, 22(12): 2200–17.

Yergeau, M.R. (2014) 'Disability hacktivism', *Computers and Composition Online*, 36, http://cconlinejournal.org/hacking/

Yusoff, K. (2019) *A Billion Black Anthropocenes or None*, Minneapolis, MN: University of Minnesota Press.

Zuckerman, E. (2017) 'Stop saying "fake news". It's not helping', [Blog] 30 January, https://ethanzuckerman.com/2017/01/30/stop-saying-fake-news-its-not-helping/

Zuckerman, E. (2021) *Mistrust: Why Losing Faith in Institutions Provides the Tools to Transform Them*, New York, NY: W.W. Norton.

# Index